WESTHILL COLLEGE OF EDUCATION

Birmingha

AF750878

A fine will be charged if this book is returned after the last date stamped below.

28 APR 1970 1 DEC 1972 9 JAN 1974 OCT 1978 30 NOV 1978 28 FEB 1979 5 DEC 1979	29 FEB 1980 6 FEB 1981 2 MAR 1981 25 MAR 1981 - 3 OCT 1995 26 FEB 1998 17 NOV 1998	28 APR 1999 25 MAR 2002	

B. L. & C.

Book No. 22485

Author Sullivan, N.

Title The message of the genes

575·1

THE MESSAGE OF THE GENES

THE MESSAGE

Navin Sullivan

OF THE GENES

ROUTLEDGE & KEGAN PAUL LTD

LONDON

First published in Great Britain 1968
by Routledge & Kegan Paul Limited
Broadway House, 68–74 Carter Lane
London, E.C.4

Printed by offset in Great Britain
in the City of Oxford at the Alden Press

© 1967 by Navin Sullivan

No part of this book may be reproduced
in any form without permission from
the publisher, except for the quotation of
brief passages in criticism

WESTHILL COLLEGE LIBRARY
22485
T1273 - 25.00
29. 10. 69
575.1 SUL

PREFACE

A new individual begins with the fusion of two microscopically tiny pieces of matter, the male and female sex cells. Together, these two cells contain all of the genes, or instructions, needed to control the correct development of a complete adult, whether plant or animal. That this can be so is one of the continuing wonders of life on earth—an event of staggering, near-miraculous complexity which is nevertheless enacted all around us every day.

Biologists have for long had to content themselves with describing the outward features of this handing on of individuality from one generation to another. They were able to work out various fairly complicated ground rules for the continuance of traits which enabled them to explain, say, why a man may suffer from hemophilia but a woman never will. However, in all of their work they dealt with, as it were, only the surface of things: they knew that something might happen but not how it happened.

To explain how a biological event happens, we must eventually understand the working of the molecules that make up all living things. Traditionally, the biologist has observed living things as functioning wholes of some kind, whereas the biochemist has studied ways in which their molecules behave in the test tube. Today, molecular biologists are combining the two approaches and are beginning to describe the functioning of living things in terms of their constituent molecules. Eventually they may be able to explain all of life in this way, and even assemble living matter. How their work has developed is explained here in relation to the greatest success of molecular biologists to date—their understanding of the message of the genes in molecular terms.

Man is, indeed, on the threshold of the mastery of life—an achievement even more profound, and more awesome, than his conquest of the atom. Let us hope that he will use his knowledge wisely.

Navin Sullivan

London
March 1967

ACKNOWLEDGMENTS

The accuracy of this book has been materially improved as a result of help that I have received from various research scientists. My particular thanks are due to Professor G. Pontecorvo, F.R.S., for his critical reading of the early chapters. Other helpful comments and information have come from many people, including Professor J. Bonner, Dr. F. H. C. Crick, F.R.S., Dr. H. Gobind Khorana, Dr. Sheila Lee, Dr. Marshall W. Nirenberg, Dr. S. R. Pelc, Professor D. C. Phillips, F.R.S., and Dr. J. Tooze.

Permission to reprint a verse by Walter de la Mare was granted by the Literary Trustees of Walter de la Mare and the Society of Authors as their representative.

CONTENTS

ILLUSTRATIONS

THE MESSAGE OF THE GENES

1

The Idea of Genes

Where did you come from, baby dear?
Out of the everywhere into the here.

Where did you get those eyes so blue?
Out of the sky as I came through.
—*George MacDonald*

WHETHER WE HAVE BLUE EYES OR BROWN IS A MATTER OF heredity. We cannot choose the color of our eyes: they are a characteristic donated to us through our parents. All physical characteristics, and most mental characteristics as well, are handed on in this way. The fifty-seven Bachs who were musicians, the thirteen Wyatts who were architects, and the eight Bernoullis who were mathematicians owed their gifts to the workings of heredity (and to training by their relatives).

Ever since people began trying to breed especially useful varieties of domestic animals and plants, they have wondered how characteristics are inherited. Without knowing, they still had considerable success—the thoroughbred race horse or the high milk yield of the modern dairy cow are evidence of this.

When Charles Darwin, the great British biologist, put forward his theory of evolution in conjunction with Alfred Russel Wallace in 1858, he realized that the workings of evolution must depend on heredity. Basically, Darwin proposed that species evolve by means of natural selection which fits them to their environments. The circumstances of any particular environment would favor individuals with certain hereditary characteristics rather than with others.

Darwin understood that continuing selection of certain characteristics at the expense of others could result, in a surprisingly short number of generations, in profound modifications to animals or plants. The success of domestic animal and plant breeders was, he saw, based on the very same principles, although the whole process was greatly accelerated by the care of the breeders.

When it came to the questions of *how* individuals come to be different, and *how* these differences recur in successive generations, Darwin had no satisfactory answers, although he did put forward a highly speculative (and incorrect) theory of his own.

Strangely enough, the first fundamental studies of heredity—studies which proved to be the essential complement to Darwin's—were going forward at the same time in a monastery at Brünn, in Moravia (now Brno, Czechoslovakia). There, a relatively obscure Augustinian monk named Johann Gregor Mendel was performing beautiful experiments in breeding peas.

Mendel had two great advantages over Darwin. He was younger, and he lived in a country where microscopes were widely used. Surprising as it now seems, Darwin was working too early to know that fertilization is the fusion of just one male cell with one female. Consequently he lacked any rational starting point from which to speculate about the mechanism of heredity. Mendel, on the other hand, used a microscope and knew that single pollen grains fertilized single egg cells. He therefore reasoned that heredity must depend on those individual germ cells and that if a plant or animal produced dissimilar offspring it must be because its germ cells were somehow dissimilar too.

From this base, Mendel began in 1856 to conduct a masterly series of experiments, aided by his mathematical training—something which many biologists yearn for even now.

INHERITANCE OF INDIVIDUAL TRAITS

When Mendel began his work the operation of heredity seemed to be full of puzzling facts. As early as 1669 Johann Joachim Becher had noted:

> When a black cock pigeon and a white hen pigeon unite, the young birds of the first generation are usually some of them entirely black and others entirely white; and it is only when we allow some of these blacks and some of these whites to unite that we get young birds which are spotted black and white. In arboriculture nature achieves similar results, for when there occurs a union between trees bearing white and red fruits respectively, spotted fruits only appear after the second crossing.

Around the time Mendel was born in 1822, an Englishman named John Goss, living in a Devonshire village, reported to the London Horticultural Society some of the strange results he had found in crossing garden peas.

Goss crossed plants yielding green peas with others yielding yellowish-white peas and in the first generation obtained only yellowish-white peas. Breeding from these hybrids, he was astonished to get three kinds of plants: one had only green peas, one only yellowish-white, and the third kind had both green and yellowish-white in the same pod.

Other people made many similar observations, without knowing how to explain them. The most notable among them was Charles Darwin, who carefully studied what happens when various varieties of garden pea are crossed. He also crossed antirrhinums, or snapdragons, using peloric and normal varieties. In the first generation he obtained only normal plants, but when he bred from these he obtained about one peloric plant to every three normal plants. However, he could not frame a satisfactory theory to account for these observations, or those made by others before him.

How did Mendel set about clarifying this confusing situation? First, he took great care in selecting his experimental material. He chose to work with the garden pea because this plant has varieties that show sharply defined differences, and because it is normally self-pollinating, so that each variety will breed true for generation after generation and not be cross-pollinated with other varieties.

Second, he selected for experiment varieties which differed in only one outward characteristic, so avoiding a confusion of data arising from simultaneous changes in many traits.

Third, he probably sat down and thought through the consequences of there being two germ cells (pollen and egg), each of which contributed to the way some particular trait was expressed in the adult formed from their union.

In 1856, Mendel crossed true-breeding varieties of pea plant that differed only in their flowers, one kind being colored and the other white. He also crossed tall plants (six to seven feet high) with dwarf plants (nine inches to one and a half feet high). To find out what the resulting hybrids would look like, he had to wait until the hybrid seeds from his cross were sown in the following year, producing another generation.

Meantime, he also made crosses involving only the character of the seeds (the peas themselves). Thus he crossed plants which produced round seeds with ones that produced wrinkled seeds, and he crossed plants that produced yellow seeds with ones that produced green seeds. Both of these yielded results in the same year, as soon as ripe seeds had been formed by the cross-pollinated plants.

To insure control of his crossings, Mendel used to open a pea flower before it was fully developed, remove its anthers (the pollen producers) by means of forceps, and then dust its stigma with pollen which he had already selected and was carrying on a camel's-hair brush. Once this was done, he tied a little paper bag around the flower to protect it from visits by any insects such as bees or pea weevils which might bring different pollen to it.

What did the first-generation hybrids show? Mendel found that those crossed between plants with colored flowers and with white flowers all grew up in 1857 with colored flowers. Similarly, those crossed between tall plants and dwarf plants all grew up tall. As for the seed characters, the plants crossed between those yielding round seeds and those yielding wrinkled seeds all produced round seeds; and those crossed between plants yielding yellow seeds and those yielding green seeds all produced yellow seeds.

In all of these experiments, the hybrids showed one trait unchanged, while the other trait disappeared. The color of the flowers was as strong as ever, the seeds were as round as ever. Evidently one trait dominated the other. Mendel named this one *dominant* and its counterpart *recessive.*

Now Mendel allowed each of these four varieties of hybrid to self-pollinate in the usual way. The generation they produced showed both traits, like Darwin's snapdragons. Thus out of the 253 hybrid seedlings from the cross between plants having round seeds and those having wrinkled seeds he obtained 7,324 seeds, of which 5,474 were round and 1,850 wrinkled. Similarly, he grew 258 hybrid seedlings from the cross between plants having green seeds and those having yellow seeds and obtained 8,023 seeds, of which 6,022 were yellow and 1,850 were green. In both of these instances, the dominant trait appeared about three times as often as the recessive. (The ratios are not exactly 3:1 for the same reason that if you dip your hand into a 3:1 mixture of red and white candies you will not bring up a handful of candies with exactly three red to every one white. Mendel realized this, which is a matter of simple statistical theory.)

Mendel found the same ratio of 3:1 held good for the proportions of plants with colored flowers to plants with white flowers, and tall plants to dwarf plants. Somehow, in all of these, the trait which had disappeared in the first hybrid generation reappeared, but only in one quarter of the total number of plants or their seeds.

INHERITANCE OF DOMINANT AND RECESSIVE FACTORS

To explain his results, Mendel proceeded as follows. Suppose the dominant hereditary factor for some particular character, flower color, is represented by *A* and the recessive hereditary factor—no flower color, or white flower—is represented by *a*. Then the germ cell from the plant with colored flowers carries factor *A*, while that from the plant with white flowers carries the factor *a*. When these two come together in pollination the result, which is called a zygote, will bear both factors, represented by *Aa*. This, then, is the constitution of the first hybrid generation.

What will happen when *Aa* factors from two hybrid plants combine in the second generation? Mendel reasoned that if each factor can join freely with any other there are four possibilities, *AA*, *aA*, *Aa*, and *aa*, each being equally likely to occur. However, as *A* is dominant and *a* is recessive,

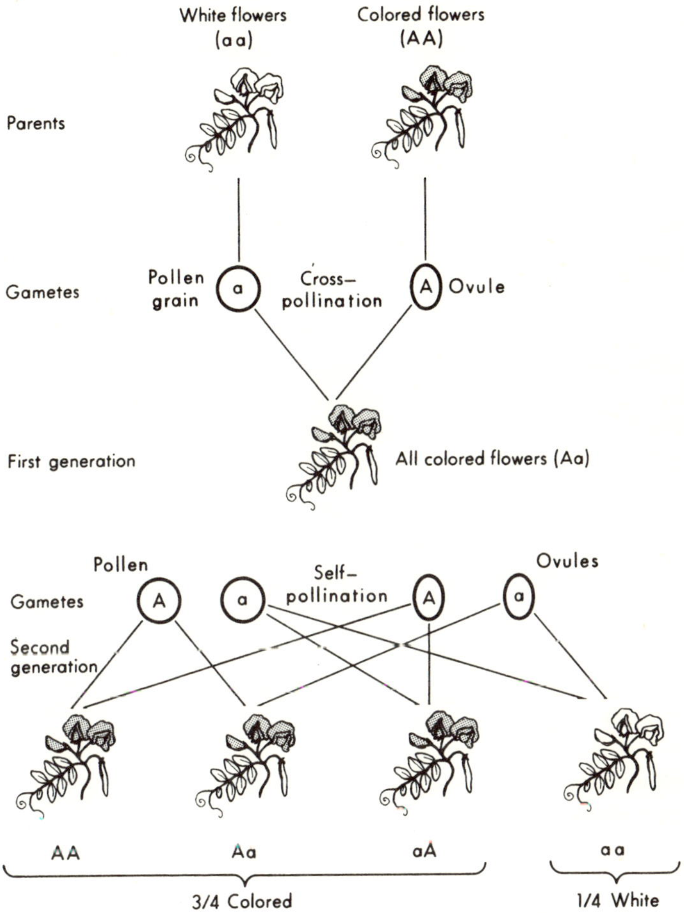

Figure 1. The result of crossing pure-breeding garden peas with colored or white flowers. Factor *A* is for flower color and factor *a* for no flower color. In the first cross these give a generation with the constitution *Aa*. From this come four possible combinations in the next generation: *AA*, *Aa*, *aA*, and *aa*. Since *A* is dominant, all plants containing it have colored flowers. Only plants with both recessive factors have colorless, or white, flowers. The overall result is the characteristic 3:1 Mendelian ratio.

flowers containing any of the first three combinations of factors will show the trait in its dominant form. Only the last possibility, *aa*, will produce flowers of the recessive form. Here, then, is the composition of the 3:1 ratio Mendel noted.

Out of these various combinations, only the flowers bearing factors *AA* or *aa* will breed true in the future. The other forms, *aA* and *Aa*, will be hybrids and in the next generation they will reveal this by once again producing dominant and recessive forms.

Mendel tested this interpretation in 1858. From 565 round seeds, products of a cross between plants yielding round seeds and plants yielding wrinkled seeds, he raised 372 plants that had both round and wrinkled seeds and 193 plants that had only round seeds. Evidently two thirds of his 565 seeds were carrying both dominant and recessive factors, while the remaining third carried only the dominant factor. This fitted the interpretation that out of every three forms carrying the *A* factor in the first generation, two were *Aa* or *aA* and one was *AA*.

To check whether traits were inherited quite independently of each other, Mendel did further experiments. In 1859 he took plants that always gave round yellow seeds and crossed them with others that always yielded wrinkled green seeds. If these four traits could be shuffled around quite independently of each other, he would get every possible combination in the first hybrid generation.

This is just what happened. From fifteen double-hybrids, Mendel obtained 556 seeds in 1860. He grew these to yield, in 1861, a total of 315 round yellow seeds, 101 wrinkled yellow seeds, 108 round green seeds, and 32 wrinkled green seeds. This corresponded very well to the ratio of 9:3:3:1 which he expected, bearing in mind the dominance of yellow over green and of round over wrinkled.

Mendel had thus demonstrated, by clear and skillfully

conducted experiments, that individual hereditary factors must exist in both dominant and recessive forms, being passed on from generation to generation in new combinations but not being changed themselves. This fundamental work, which forms the basis of all modern genetics, had occupied him for eight years.

One evening in February 1865, Mendel and some of his friends set out for the Brünn Society for the Study of Natural Science, where he was to announce his results. We can picture the stocky stout man wearing his teacher's garb of tall hat, long black coat, and trousers tucked into top boots, striding through the streets with his precious manuscript under his arm. His excitement must have been intense.

About forty persons heard Mendel that evening as he described the ratios he had obtained. They also duly turned up again the next month to hear his explanation of them. However, the significance of his work completely escaped them all. It made so little impact that there was not even any discussion! We can only guess at the depth of Mendel's disappointment.

Nevertheless, his lectures were printed in the official proceedings of the society and were sent, as was customary, to more than 120 other learned societies, universities, and academies all over the civilized world. That did not help: no one understood Mendel's work, and the proceedings of the obscure Brünn Society merely lay on the shelves, gathering dust. Even as late as 1884, when Mendel died, his work remained unknown. Why was this?

The most probable reason is that the scientific world was still busily digesting Darwin's epoch-making work on evolution. It was not yet ready for Mendel, although ironically enough Mendel's work was needed to explain how Darwin's ideas of natural selection could operate. Like many pioneers, he was so far ahead of his time that he died unrecognized.

Today we count him the father of heredity. Indeed, some go even further and call him the father of molecular biology.

Actually, Mendel was lucky. In all, he studied seven hereditary traits. Each showed a clear-cut pattern of dominance and recessiveness, and each could be inherited quite separately from any of the others. This is not always true. For instance, in shorthorn cattle the crossing of a red bull with a white cow yields not all red offspring, or all white offspring (as would be the case if one of these traits were dominant), but a combination of red and white which we call roan. Linkage, too, can exist between hereditary factors so that inheritance of one brings with it inheritance of another.

By pure chance, it happened that each of the seven characters Mendel studied can, in the pea, be inherited independently of the other six. Further, this is the largest number of characters in the pea that can be inherited quite independently of each other! Truly, fate was on Mendel's side.

REDISCOVERY OF MENDELIAN LAWS

Sixteen years after Mendel's death, his work was rescued from oblivion in one of the most dramatic rediscoveries in the history of science.

Hugo de Vries, in Holland, published an account in March 1900 of very extensive research which he had performed on plant hybrids. He put forward two propositions which, he said, "were, in essentials, formulated long ago by Mendel for a special case. . . . They fell into oblivion, however, and were misunderstood. According to my own experiments, they are generally valid for true hybrids." De Vries added that he had not become acquainted with Mendel's work until he had independently reached the same conclusions.

A mere month later Carl Correns, working at Tübingen

in Germany, wrote that he, like de Vries, had independently worked out the ideas of dominant and recessive traits, and of the separation of the factors for these pairs in the sex cells. Subsequently, he had discovered that Mendel had anticipated him. However, he went further than Mendel, because he found that "in the case of a great many pairs of characters, we do not find that either of them is dominant."

As though this were not enough, in June of the same year a Viennese botanist, Erich Tschermak, set out the results of his own research on peas, following upon Darwin's earlier observations. He, too, had independently arrived at the same conclusions as Mendel. Gallantly, he added a postscript to his paper: "The simultaneous discovery of Mendel by Correns, de Vries, and myself seems to me peculiarly gratifying. I, too, as late as the second year of my experiments, believed that I had happened upon something new."

GENES AS MATERIAL PARTICLES

William Bateson, an English scientist, read about this rediscovery of Mendel's work while traveling by train from Cambridge to London. At once his imagination was fired. For the rest of his life he was a staunch supporter of Mendelian ideas, and he interpreted them to the next generation. He also invented the name "genetics" for the whole study of heredity.

Curiously enough, however, Bateson never wholly accepted Mendel's findings. He was enthusiastic about the fact that individual traits were inherited, and were segregated from each other in the sex cells, yet he balked at the idea that particles of matter could carry these on from one generation to another. "Genetic factors are definite things," he wrote in 1914, "either present or absent from any germ cell." Yet he had also said, a paragraph earlier, that "it is unlikely that the genetic factors are in any simple or literal sense material particles."

Bateson could not accept the idea that the inherited characteristics of himself and other people might stem from nothing more than a collection of molecules. A Danish botanist, Wilhelm Johannsen, was equally uncertain. Although he decided that definite units in the sex cells must determine the hereditary characteristics, and christened them "genes," he did not want to allocate any material basis to them.

In spite of this, it became increasingly evident to scientists that genes must be real things. In that case, where were they located?

To answer this question required the help of cell scientists, or cytologists–men who peered down the microscope at that smallest unit of life, the individual cell.

2

Chromosomes and Genes

THE EXISTENCE OF CELLS WAS FIRST NOTED BY AN ENGLISHMAN, Robert Hooke. In 1665 he wrote a description of his findings:

I took a good clear piece of cork and with a penknife sharpened as keen as a razor, I cut a piece of it off, and thereby left the surface of it exceedingly smooth, then examining it very diligently with a microscope, methought I could perceive it to appear a little porous; but I could not so plainly distinguish them as to be sure that they were pores . . . I with the same sharp penknife cut off from the former smooth surface an exceeding thin piece of it, and placing it on a black object plate . . . and casting the light on it with a deep plano-convex glass, I could exceedingly plainly perceive it to be all perforated and porous, much like a honeycomb, but that the pores of it were not regular . . . these pores, or cells, were not very deep, but consisted of a great many little boxes, separated out of one continued long pore by certain diaphragms . . .

Nor is this kind of texture peculiar to cork only; for upon examination with my microscope, I have found that the pith of an elder, or almost any other tree, the inner pulp or pith of the cany hollow stalks of several other vegetables: as of fennel, carrots, daucus, burdocks, teasels, fern, some kinds of reeds, etc., have much such a kind of schematisme as I have lately shown that of cork.

What Hooke was describing was the outer boundary, or cell wall, of the cell in certain plants. Animal cells have no such strong walls. Both plant and animal cells are alike, however, in containing living material, the cytoplasm. In the cytoplasm is the nucleus, which is the controlling center of the cell.

The nucleus was first seen by Antony van Leeuwenhoek, a draper in Holland who spent his spare time grinding lenses and fashioning the best microscopes of his day so that he might discover a new world, hidden from the naked eye. In 1700 he spread a drop of salmon's blood on a very clean glass and saw the little oval particles—blood corpuscles, as we now call them—that floated in it. He noticed that in some of these particles there was "a little clear sort of light in the middle." Without knowing it, he had seen the nucleus of a living cell.

Then, in 1781, Felix Fontana found oval bodies inside skin cells from an eel, so making the first observation of a nucleus in an adult animal cell other than a blood corpuscle. Once again, however, the nucleus was described but not recognized for what it was.

The man who established the idea that a nucleus is a normal component of the living cell was a Scottish botanist, Robert Brown. In 1883 he wrote that he recognized a nucleus in the cells of a number of flowering plants, and gradually he found that a nucleated plant cell was normal.

Observations on animal cells were harder, because of the bewildering diversity of their shape and size. Many workers contributed to the idea that animal cells, too, have a nucleus.

CELL DIVISION

In 1839 two German scientists, Theodor Schwann and Matthias Jakob Schleiden, a zoologist and a botanist respectively, took many scattered ideas and observations and welded them together. The cell, they suggested, is the basic unit of life in both plants and animals.

Some twenty years later Rudolph Virchow, a great German physician, went a step further and stated that cells only come from pre-existing cells. He wrote:

> Where a cell exists there must have been a pre-existing cell, just as the animal arises only from an animal and the plant only from a plant. The principle is thus established, even though the strict proof has not yet been produced for every detail, that through the whole series of living forms, whether entire animal or plant organisms or their component parts, there are rules of eternal law and continuous development, that is, of continuous reproduction.

The division of plant cells was brilliantly studied by a self-taught amateur, Wilhelm Friedrich Hofmeister. While working in his father's bookselling and publishing business in Leipzig, Germany, he emulated Leeuwenhoek in spending all his spare hours at the microscope. He even rose at four o'clock in the summer mornings so as to get in two hours' work before going to the office. Eventually, after publishing many papers in the 1840s and 1850s, he was offered an academic post, and in 1863 he became professor of botany at Heidelberg University.

Hofmeister carefully observed that, before a cell divides to produce two new cells, its nucleus divides to produce two daughter nuclei. He also recorded that the nucleus first resolves itself into smaller bodies—bodies which, after cell staining with aniline dyes became possible, were called chromosomes (from the Greek meaning "colored body") because they take up certain dyes readily.

Eventually scientists found that the nucleus of any plant or animal cell, when about to divide, resolves itself into rod-shaped chromosomes. Further, the number of these chromosomes is always the same in each animal or plant, although it varies widely from one type to another. Thus ordinary human body cells have forty-six chromosomes, those of maize have twenty, the mouse has forty. The lowest number known is two, in the roundworm, while amoebas and some plants have hundreds of chromosomes.

MITOSIS

Between 1875 and 1880 scientists generally came to agree that cells in both plants and animals are formed by equal division, with the nucleus always dividing before the cell. Initially there was confusion about the chromosomes, some workers maintaining that these divided in half across the middle. However, the German scientist Walther Flemming, using various aniline dyes to stain the dividing cells, saw as early as 1879 that the chromosomes divided in half along their length. The two halves then separated, one going to each daughter nucleus, which thus became a replica of the parent.

By 1882 Flemming had distinguished nine phases in this process of chromosome division—a process which he christened mitosis. The way we portray mitosis today is shown in the accompanying diagram.

First, a small body near the nucleus called the centriole divides and separates, throwing out what seems to be a radiating system of protein fibers. These two radiating systems, the asters, travel to opposite poles of the cell and then become connected by fibers which form a system called the spindle.

Meanwhile the chromosomes, which are invisible in the resting nucleus, arrange themselves along the equator of the spindle, midway between the two centrioles. Each chromosome is split longitudinally to become twin threads, joined only at one place called the centromere. After arranging themselves at the equator, the chromosomes divide completely.

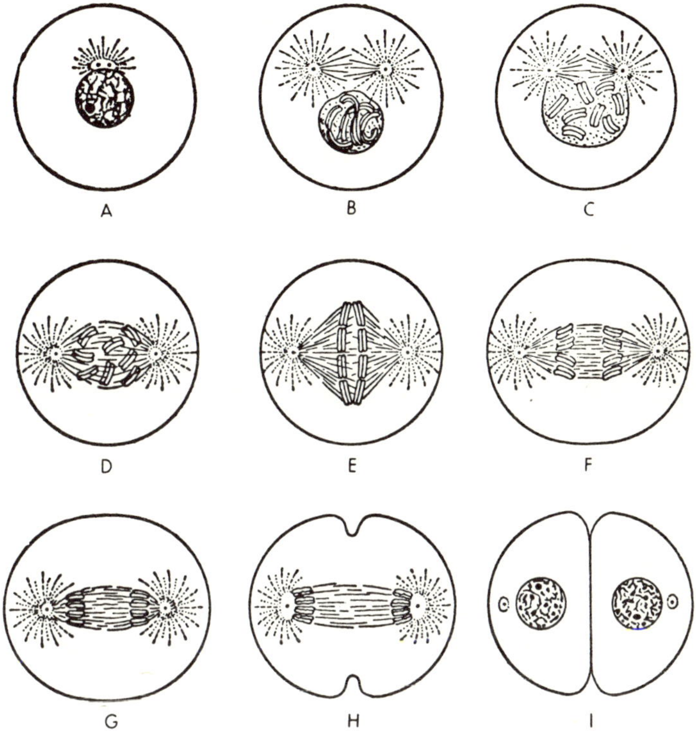

Figure 2. The progress of mitosis, in which a cell divides to form two identical daughter cells, each with the same complement of chromosomes as the parent.

Now one complete set of daughter chromosomes migrates to each pole and in coordination with this the entire cell divides, animal cells by pinching in two, plant cells by building a new wall. The daughter chromosomes disappear as a new resting nucleus forms in each daughter cell.

This whole process, which resembles some stately minuet, can occur in as little as half an hour and will normally take no more than a few hours. It is essentially the same in all animals and plants.

FERTILIZATION

Mitosis is the method by which a single cell divides in the body of an animal or the structure of a plant. In some marvelous way, each daughter cell is a replica of its parent. However, a still more marvelous process of generation occurs when not single cells but whole organisms reproduce. Here is the problem of passing on hereditary information at its most fantastic. Not only must new cells be successfully made, but they must be grouped together into a multicellular whole, complete with specially differentiated structures such as sense organs and limbs in a mammal or leaves and roots in a flowering plant.

The starting point for the growth of an entire new organism is the fusion of a single male cell and a single female cell. In a human, the male cell, or spermatozoon, is far too small to be seen with the naked eye. First described by Antony van Leeuwenhoek, it is a tadpole-shaped structure so small that forty million sperm—enough to found a medium-sized country—would take up no more room than a single pinhead. In comparison, the human female cell, or egg, is enormous, and is just visible to the naked eye. It was first conjectured by a contemporary and fellow townsman of Leeuwenhoek's, Regnier de Graaf, from his observations of the ovary.

That two such minute lumps of matter can contain all the genetic information needed to construct a human being is near-miraculous. Not surprisingly, the men who first considered the idea tried to invent simpler-seeming explanations. Unaware that egg and sperm fused—unsure, indeed, that both were necessary in reproduction—they credited "explanations" in which one of these carried a miniature model of the parent.

Marcello Malpighi, an Italian, looked at a fertilized hen's egg with its developing embryo and suggested that the egg carried this model, just as a flower bud contains the flower preformed. Others, such as Jan Swammerdam in Holland, imagined that their microscopes revealed a tiny embryo in the human spermatozoon.

One man who tried to resolve this sperm versus egg controversy was an Italian abbot named Lazzaro Spallanzani. He believed that it was the egg which carried a miniature embryo. To decide between sperm and egg, he took frogs and toads at mating time and fitted them out with silk pants. As a result, he found that the eggs shed by the females remained sterile. Then he strained the seminal fluid shed by the males through filter paper and tried its effect on the eggs. Nothing happened. By these observations, Spallanzani had of course demonstrated that sperms are needed to fertilize eggs—but he did not want to believe this and, being wily and determined, succeeded in not doing so.

Not until 1854 was a sperm actually seen making its way into the egg of a frog, and only in the following year was the same process seen in various pond weeds.

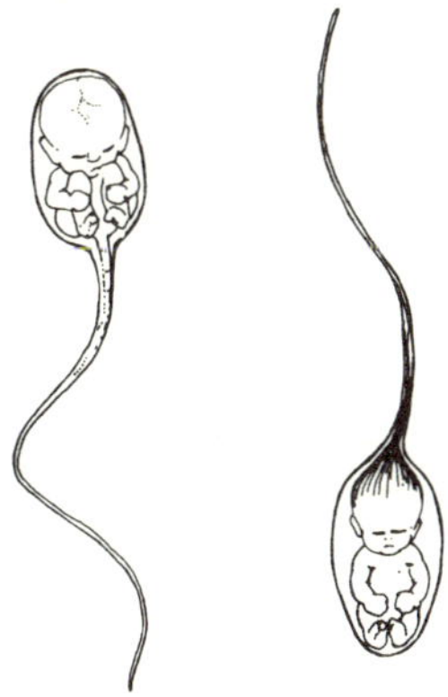

Figure 3. Ancient microscopists imagined that the human sperm contained tiny embryos.

The sperm consists of very little but nucleus, with a propulsive tail. What happens to this nucleus in the moment of fertilization?

Even when the basic idea of fusion between egg and sperm had been accepted, some scientists thought that the job of the sperm was merely to enter the egg and, by penetrating it, stimulate it to divide. However, in 1875 Oscar Hertwig in Munich, Germany, proved that the nuclei of the sea urchin egg and sperm fused, and that only one sperm reached its goal. Two years later the same was demonstrated for plants by Eduard Strasburger, working with the lily.

These discoveries about the moment of fertilization were revolutionary. They demonstrated that a new individual begins at the moment of fusion between the nuclei of egg and sperm. From then on, the organism grows by continued cell division and cell differentiation.

MEIOSIS

The number of chromosomes in the nucleus is normally constant for any species. However, the sex cells do not have the same number of chromosomes as the body cells. If they did, a human egg and sperm each containing 46 chromosomes would unite to form a zygote containing 92 chromosomes. If this process continued, by the end of the tenth generation the individual would have cells each containing 23,332 chromosomes.

To avoid this ridiculous state of affairs, a special kind of cell division occurs in the formation of the sex cells, during which the number of chromosomes is reduced to half. Thus the normal complement is made up when the two sex nuclei fuse. This process can occur only because in the normal cell the chromosomes come in pairs: in the sex cells only one member of each pair is present.

This process of reduction division was first suggested by a German zoologist, August Weismann, while he was attending a meeting of the British Association for the Advancement of Science at Manchester in 1887. He put the idea forward to explain why chromosome numbers stayed constant from generation to generation.

Five years later, Weismann made another very important suggestion. This was to account for the fact that different offspring from the same parents are not identical. He proposed that, since each offspring must have different hereditary characteristics, these characteristics are reshuffled in the formation of the parents' sex cells. This would mean that different sex cells in one organism would contain different combinations of hereditary characteristics. He added that this reshuffling must occur during reduction division.

During reduction division, elaborate chromosome maneuvers occur. In the early stages, each member of a chromosome pair moves toward its fellow, until each pair is united along its whole length. By some marvelous procedure, each bit of discernible detail on one chromosome comes exactly opposite the equivalent on the other chromosome. When the chromosomes are closely alongside each other in this way, the nucleus looks as though it has only half its normal complement of chromosomes.

The chromosomes now shorten and thicken and divide in two longitudinally, though this is not evident until the next stage, when the two pairs thus formed begin to separate. Separation is not, however, complete because at one or more points along their length the members of each double chromosome remain in contact. Consequently odd shapes are produced as the chromosomes cross over each other. These contact points, or crossings-over, were christened chiasmata (singular chiasma, from the Greek *chi* [χ]) by Frans-Alfons Janssens, a Jesuit professor at the University of Louvain in Belgium, working in 1909. We now know that some exchange of chromosome material between the pairs occurs at this stage.

Next, the double chromosomes separate from each other

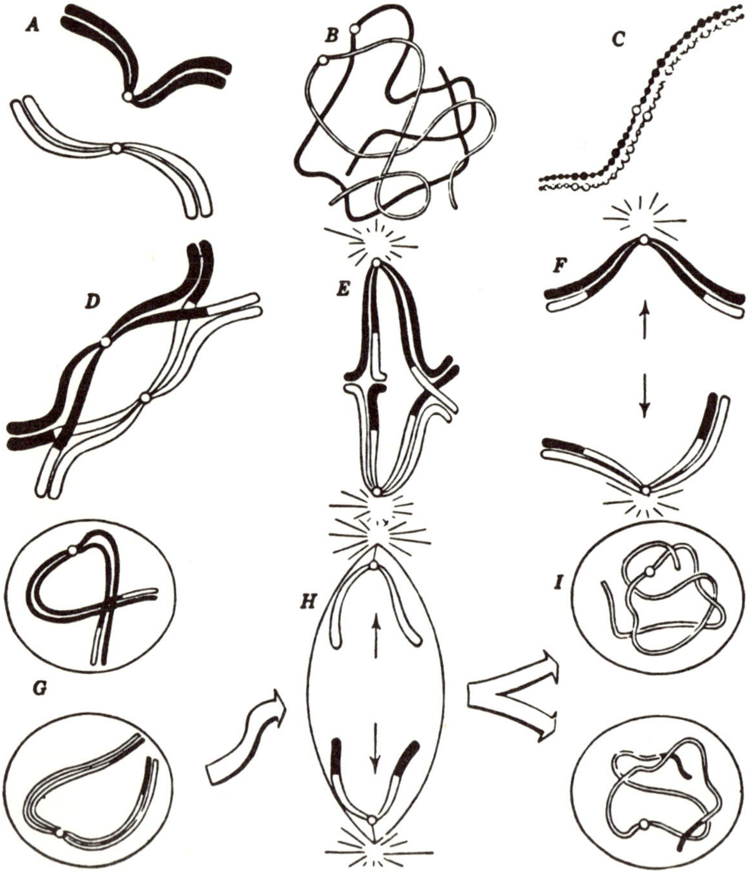

Figure 4. The progress of meiosis, or reduction division. The centromeres are shown as white circles.

and move to opposite poles of the cell. A nucleus then forms at each pole, and the cell divides in two. At this point there are two daughter cells, each containing double chromosomes. The number of double chromosomes in each daughter cell is half the normal chromosome number for the organism.

Finally, each of these double chromosomes moves apart and each daughter cell divides in two. The result is four cells in all, each with half the normal number of chromosomes. These are the sex cells.

The total process of reduction division is termed meiosis. The elaborate chromosome maneuvers that occur during it are a distinctive feature and they suggest a recombination of material of the kind envisaged by Weismann. If hereditary characteristics are carried on chromosomes, they would be reshuffled during every meiosis.

CHROMOSOMES AND GENES

The likelihood that hereditary factors, or genes, were carried on chromosomes was in fact argued in 1902 and 1903 by Walter S. Sutton, a graduate student working at Columbia University under the great cell scientist Edmund Beecher Wilson.

Sutton pointed out that genes must be located in the egg and sperm cells since these, and these alone, carried hereditary material from one generation to the next. Further, the nucleus is the most significant part of the sperm and is similar to the nucleus of the egg. Since egg and sperm make equal contributions to the heredity of the zygote, it is reasonable to suppose that they do so through their nuclei. What bodies in the nucleus might bear the hereditary factors?

In answer, Sutton detailed the correspondence between the known movements of the chromosomes and the assumed movements of the genes as required by Mendelian laws.

1. Egg and sperm each carry half the number of chromosomes (and genes) found in body cells.
2. In meiosis the individual pairs of chromosomes separate. Genes, too, separate by pairs, so that *AA*, *Aa*, or *aa* combinations are separated into *A* and *A*, or *A* and *a*, or *a* and *a*.
3. Formation of the zygote brings the number of chromosomes (and genes) up to the number in the body cells. If the sex cells each carry a chromosome with factor *A* or *a*, the hereditary composition of the zygote will follow the rules proposed by Mendel.
4. Chromosomes retain their existence throughout cell divisions, and so do individual genes.

From these points Sutton came to a general conclusion:

> Thus the phenomena of germ-cell divisions and of heredity are seen to have the same essential features, viz. purity of units (chromosomes, characters) and the independent transmission of the same; while, as a corollary, it follows in each case that each of the two antagonistic units (chromosomes, characters) is contained by exactly half the gametes (sex cells) required.

When making the investigations which led him to this statement, Sutton had observed chromosome movements in a particularly favorable species, a type of grasshopper bearing easily distinguishable chromosomes in its sperm-forming cells. His verdict, that genes were probably borne by chromosomes, revealed the true significance of all the complicated processes of meiosis which men had already observed through their microscopes. In this, Sutton achieved a decisive step in the progress of biology.

However, Sutton had not proved that chromosomes are the bearers of genes: he had only demonstrated the likelihood that they are. Proof was to be brought nearer by another researcher at Columbia University, Thomas Hunt Morgan (a relative of J. Pierpont Morgan, the financier).

SEX-LINKED GENES

Sutton had pointed out that there are more hereditary factors than there are chromosomes in any species. Consequently, he argued, one chromosome must bear a number of factors, which in this way would be linked together. In 1905 this idea was borne out by two British researchers, William Bateson and Reginald Crundall Punnett, who found in experiments on sweetpeas that the factors for blue flowers and long pollen grains can be associated together, or linked.

Morgan was to discover a similar phenomenon with fruit flies. These tiny black flies, which are commonly found on overripe bananas left lying around during the summer, grow from egg to maturity in only twelve days and are therefore good material for displaying hereditary changes quickly in succeeding generations. Their scientific name is *Drosophila melanogaster* (meaning "black-bellied dew lover"!). Clyde Melvyn Woodworth, later professor of entomology at the University of California, was the first to demonstrate the usefulness of these flies as laboratory animals.

After visiting Hugo de Vries' experimental garden in Holland around 1900, Morgan had been greatly impressed by the way in which numerous sharply distinct types of flower could be bred—types that differed from their parental forms and went on breeding true to their new characteristics. Such sharply changed, true-breeding types are called mutants.

Around 1909, Morgan began trying to induce mutations in *Drosophila*. To produce these, he tried a wide range of temperature, salts, sugars, acids and alkalies, and radium and X rays. The last two should have been effective but Morgan's technique for detecting mutations was inadequate, and those mutations which he did find were not attributed to his treatment.

After a year of breeding a pedigree culture of *Drosophila* through many generations, Morgan noticed an extraordinary happening. Among all the hundreds of normal, red-eyed flies there had suddenly appeared a male fly with white, or albino, eyes! Here was a mutation of the very kind he had been hoping for.

Morgan bred the white-eyed male to his red-eyed sisters and obtained 1,237 offspring. All of these had red eyes, which was not surprising as red eyes are normal and therefore presumably dominant to white. Now he crossed these red-eyed offspring among each other, expecting to obtain a classic Mendelian ratio of three red-eyed flies to one white-eyed. In fact, he obtained 3,470 red-eyed flies to 782 white-eyed ones—and every one of those white-eyed flies was male!

In later generations, white-eyed female flies appeared among Morgan's collection and so he was able to do a switch. He bred a white-eyed female to a red-eyed male and found that all the daughters were red-eyed and all the males were white-eyed.

Clearly, the gene for white eyes was somehow linked to the mechanism of determination of sex. But how could this be? Careful work on the microscopic appearance of the cells of *Drosophila* pointed to an answer.

Out of the four pairs of chromosomes in an ordinary body cell, three have similar members in both sexes; the fourth pair, however, is composed of two straight chromosomes (called *X* chromosomes) in the female, and an *X* chromosome and a hook-shaped *Y* chromosome in the male. Here, then, was an important indication that chromosomes were the actual carriers of hereditary factors.

Further, Morgan was able to explain the way in which white eyes were inherited by supposing that a gene for whiteness could be carried on an *X* chromosome but never on a *Y* chromosome. As it was a recessive gene, it appeared only when it was not masked by the dominant gene for red eyes, which also could be carried on an *X* chromosome. Thus the recessive gene could appear most easily in the male, where the *Y* chromosome had no gene to mask it, and it could also appear in the female if both *X* chromosomes carried it.

Morgan made further experiments to find out how often

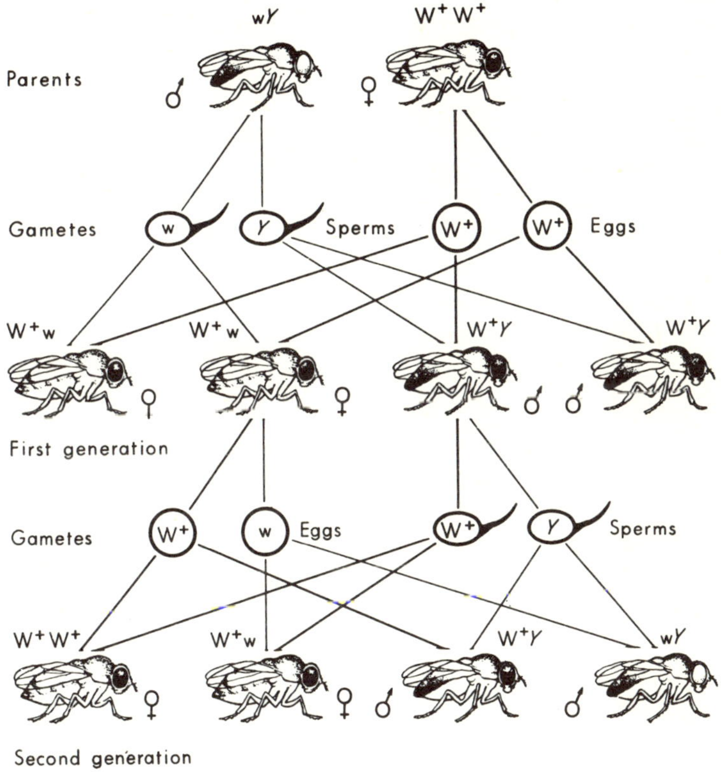

Figure 5. The inheritance of white eyes in *Drosophila* is a sex-linked phenomenon. The gene for whiteness, *w*, is recessive and carried on the *X* chromosome. It most easily appears in the male because the *Y* chromosome of the male has no dominant *W* gene for red eyes to mask it.

traits were inherited in groups instead of quite separately according to simple Mendelian laws. He discovered that certain traits normally appeared together. For instance, a fly with a black body would have tiny vestigial wings and a fly with a "gray" body (actually, a yellowish body with black bands) would have full-sized wings.

If these two types of fly were crossed, the usual outcome was flies with the same combination of characteristics as in the parents, not a fresh assortment of "gray" body and vestigial wings, or black body and full-sized wings. Evidently the traits for "gray" body and full-sized wings were linked; so were those for black body and vestigial wings.

Proceeding in this way, Morgan studied the inheritance of hundreds of traits in *Drosophila*. He bred the little fruit flies in the thousands, feeding them on banana mash and keeping them in milk bottles labeled with any odd scraps of paper —even discarded envelopes out of his mail. Every available inch in his "fly room" was occupied by a milk bottle carrying its particular crew of flies. Summers, barrel-loads of these bottles and their precious occupants were shipped to Woods Hole, Massachusetts, where Morgan and his collaborators could go on working away from the heat and humidity of New York City. No one before had studied inheritance on such enormous numbers of individuals—and laborious work it was too, for every tiny fly had to be studied under a hand lens.

However, Morgan's thoroughness paid off. He was able to determine that *Drosophila*'s individual traits are linked to each other in four groups. Morgan also knew that there are just four pairs of chromosomes in the body cells of this fly. Definitely, it looked as though the genes for these traits were located on the chromosomes.

The work that finally clinched the fact that chromosomes are indeed bearers of hereditary material was published in 1916 by a graduate student of Morgan's named Calvin B.

Bridges. He was studying the way that a gene for specially bright red eyes, called vermilion, was transmitted in *Drosophila*. It turned out to be a recessive gene which, like the gene for white eyes, seemed associated with the *X* chromosome. This meant that a cross between vermilion-eyed females and normal-eyed males would usually give rise to normal-eyed females and vermilion-eyed males.

Very occasionally, however, the seemingly impossible happened, and this cross produced a vermilion-eyed female! Somehow, if the theory of recessive genes were correct, this female must have inherited *two* recessive genes for vermilion —and yet the gene received from her normal-eyed father would not be recessive. Where, then, could the two recessive genes have come from?

Bridges had to decide between abandoning current genetic theory or devising an unorthodox explanation to account for the transmission of those two recessive genes. He chose the unorthodox and guessed—brilliantly—that the mother fly's two *X* chromosomes might have failed to separate at meiosis, so being transmitted as a pair to the daughter, along with the usual single sex chromosome from the father fly. On examining the cells of these exceptional daughter flies under the microscope, Bridges found that his hunch was correct. There were three sex chromosomes instead of the normal two, one being a *Y* and the other two *X*.

The chromosome theory of heredity was finally established.

MAPPING THE CHROMOSOMES

Although scientists now knew that the genes of *Drosophila* were carried in four groups on four pairs of chromosomes, they did not know the way in which those genes were placed.

As early as 1911, they wondered about the possibility of "mapping" the chromosomes to show the relative positions of the genes.

In meiosis, one sequence in the chromosome maneuvers involves crossings-over between double pairs of chromosomes, described by Janssens in 1909. Morgan realized that these crossings-over could provide a way of deducing the order of genes in each chromosome. He argued as follows.

Suppose the genes are arranged in line along each chromosome. Then, when crossing over occurs, a gene may be transferred to another chromosome. If one chromosome were composed of genes *ABCDEFGHIJKL* and the other one in the pair had corresponding genes *abcdefghijkl*, they could twist together and interchange portions so that the resulting chromosomes were *ABCDEFghijkl* and *abcdefGHIJKL*. This kind of process is termed recombination, as the genes in a pair of corresponding chromosomes are recombined in the crossing-over process.

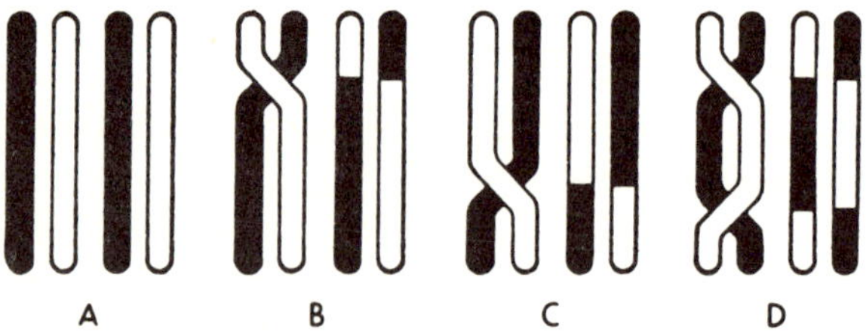

Figure 6. As a consequence of crossing over between double pairs of chromosomes, which occurs during meiosis, corresponding parts may be interchanged between chromosomes, which then are said to undergo recombination.

Now suppose that two particular genes are at opposite ends of a chromosome. When that chromosome breaks in crossing over both the genes may be interchanged, but it is highly likely that only one will be interchanged and so they will be separated to different chromosomes. If, however, the genes are located nearer together, separation is less likely. The likelihood that any two genes will be separated from each other in recombination is thus closely related to their relative positions on a chromosome. Because of this, Morgan argued that it should be possible to plot the sequence of genes in a chromosome by analyzing statistically how often any two of them were separated during recombination.

By patiently analyzing many matings between fruit flies bearing particular mutations, Morgan and his students Alfred H. Sturtevant, Calvin B. Bridges, and Hermann Joseph Muller were indeed able to place these mutation sites in linear order. By 1915 they had published a preliminary map showing the relative locations of altogether fifty genes on the four *Drosophila* chromosomes. Later work has increased this number enormously: by now, one of this fruit fly's chromosomes alone is known to contain some five hundred genes. Similar maps have been constructed for many other organisms as well.

Thus far, the chromosome maps for *Drosophila* were statistical charts. Although they enabled geneticists to prophesy with uncanny accuracy the distribution of characters among the descendants of a cross, they were not actual maps showing the physical locations of the various genes. How intriguing it would be, thought geneticists, if it were possible to magnify chromosomes enough to detect any physical differences that might identify individual genes! However, chromosomes are so small that they did not expect this would ever be possible.

In 1930 D. Kostoff pointed out that the geneticists' desires might, after all, be fulfilled. The salivary glands of *Drosophila* contain giant-sized chromosomes, some one hundred to two hundred times longer than ordinary chromosomes. They were first described by a Frenchman, E. G. Balbiani, in 1881, but geneticists had overlooked their existence. When stained and squashed on a slide, these chromosomes show thousands of cross-bands all the way along them. Kostoff suggested that these bands might correspond with the arrangement of the genes.

From 1933 onward, Theophilus Shickel Painter of the University of Texas began an intensive study of these chromosomes. In his first paper he reported:

> The same chromosomes, or characteristic parts thereof, may easily be recognized in different cells . . . This discovery places in our hands, for the first time, a qualitative method of chromosome analysis . . . by studying chromosome rearrangements of known genetic character, we can . . . construct chromosome maps with far greater exactness than has been heretofore possible.

Painstaking work comparing the changes in appearance of these chromosomes with the genetic changes in their owner led to maps of the physical position of the genes. These physical

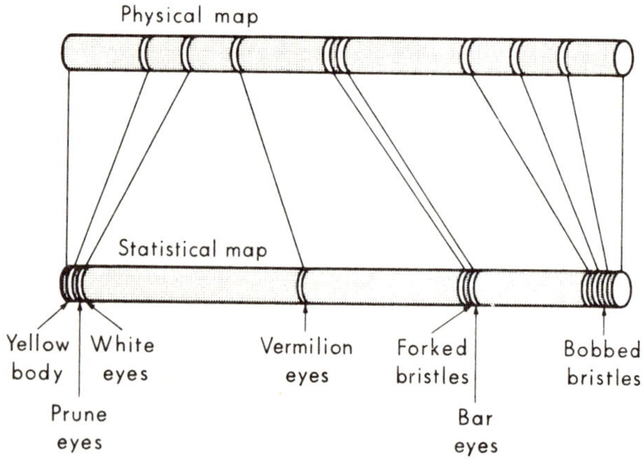

Figure 7. Statistical and physical maps for the positions of genes along a chromosome do not coincide exactly because the likelihood of crossing over per unit length is not the same everywhere along the chromosome.

maps did not look like the statistical ones, although the order of the genes was the same. One reason was that some parts of the chromosome are more liable to cross over than are others, and so are farther away from each other on the statistical map than they are in reality.

The mapping of chromosomes showed beyond doubt that genes are arranged in line along chromosomes. Growing still more ambitious, researchers now endeavored to estimate the size of an individual gene. Thus in 1934 H. J. Muller and A. A. Prokofyeva reported from recombination studies that the maximum length of a single gene in *Drosophila* was 1,250 angstroms (one angstrom equals one hundred-millionth of a centimeter).

WHEN IS A GENE A GENE?

By this time it was becoming apparent that a gene was a pretty complicated idea, varying in definition with the way that a researcher wanted to regard it. Thus it could be defined as the basic unit of self-reproduction, or the basic unit involved in a mutation, or the basic unit affecting the physiological character of the organism in some way, while from a purely genetic point of view it was the ultimate unit involved in recombination. It turned out that these definitions did not always mean the same thing.

For instance, the size of a gene as defined from recombination studies seemed to be about 1,000 angstroms. On the other hand, by using X rays to induce mutations and calculating the area hit by the X-ray beam, researchers figured that irradiating an area only 100 angstroms across could be enough to cause a mutation. Should this target area be thought of as a gene? If so, a gene defined in this way was a tenth of the size of one defined through recombination studies.

Still more complicated was the finding that the effectiveness of a gene in influencing the characteristics of an organism depended on its relation to its neighboring genes. This so-called position effect was closely studied by Muller, who suggested that gene functioning might partly depend on overlapping regions between neighboring genes. If this were to be so, the exact limits of an individual gene would constantly be shifting, depending upon its neighbors at each particular combination.

In 1940 Muller and his colleague Daniel Raffel wrote: "The idea of a 'whole gene' or, more directly, of 'a gene' itself, thus comes under scrutiny." They also made a highly significant suggestion that the "gene," in the rather loose sense so far used by geneticists, might be divided into still smaller pieces which under some conditions could act as independent units—for example, in recombination.

This idea was followed up by Guido Pontecorvo, working at the University of Glasgow in Scotland. In 1952 he suggested that the ultimate units involved in crossing over and in mutation might not only be different from each other but be only small portions of the chromosome segment constituting a physiological gene—that is, the unit governing some definite physiological characteristic of the organism.

Exploring subdivisions of the gene still further, Pontecorvo drew attention to the fact that several mutation sites situated close together might produce different effects, depending on whether they were all on one member of a chromosome pair or distributed between both members.

One such case involved the ability of the mold *Aspergillus nidulans* to synthesize a substance called biotin. This ability could be destroyed by mutation at any one of three closely grouped sites on the chromosome. If any two of these mutant sites were on only one member of a chromosome pair, their effect (which was that of a recessive hereditary factor) was masked by the other, normal chromosome which carried the dominant factor. If, however, one mutant site was on one member of a chromosome pair and another was on the second member, they took effect and the mold starved to death.

How was this to be interpreted? Pontecorvo spelled out two possibilities: either three separate genes were involved or there were three mutant sites within one gene. If, as he considered, the second were the likelier explanation, this meant that the gene could be analyzed into three constituent parts, a mutation in any one of which would be enough to inactivate the gene as a whole. Thus the presence of any one of the three mutants on one chromosome, and of another on the other chromosome, would put the gene out of action on both.

FINE GENETIC ANALYSIS

We now know that the second explanation is correct. In proposing it, Pontecorvo was pointing the way to fine genetic analysis—the analysis of sites within a single gene. Since then, this work of fine genetic analysis has been carried much further, notably by the massive research effort of Seymour Benzer and his associates at Purdue University.

The main snag with mapping mutation sites within a gene is that the sites are so close together. As the chance of sites being separated during recombination gets less the closer together the sites are on the chromosome, this means that mutation sites within a gene are only rarely separated. It is therefore necessary to search through millions of individuals, on the average, to find one in which site separation has occurred during recombination.

With fruit flies, the task is impossible as the researcher deals with at most a few thousand individuals and each generation takes about twenty days. With bacteria, or with a virus of the bacteriophage type which infects bacteria, the situation is much better. Here the researcher is dealing with millions of individuals and a generation takes only minutes to grow. Thus a test-tube experiment taking twenty minutes can yield genetic data which, if humans were used, would require the entire population of the earth!

This magnifying effect in genetic analysis has been brilliantly exploited by Benzer, using a bacteria-infecting virus, or bacteriophage, designated T4. One classic piece of work was done by him on a class of T4 mutants called *r*II. When the virus is used to infect two different strains of *Escherichia coli* bacteria, the B and K strains, *r*II mutants are distinguished by the fact that they grow in the B strain but not the K strain.

Benzer found that if he infected K strain bacteria with two different *r*II mutants, some of the viruses would, after all, grow vigorously. The reason for this turned out to hinge upon the way in which genetic material from the two mutants could recombine while they were together in a single bacterial cell. Suitable recombination of the material could lead to the production of a reconstituted normal-type virus.

This result held good only for certain pairs of *r*II mutants. By fine mapping techniques, Benzer was able to locate the mutation sites on these mutants. He found that they always lay toward opposite ends of the *r*II region of the genetic map. He could thus divide this region into two parts, which he christened cistrons, designating one A and the other B.

If an *r*II mutant with a mutation in the A cistron was recombined with another bearing its mutation in the B cistron, it was possible for the result to be a virus made up of a normal B cistron plus a normal A cistron. On the other hand, if both mutants had their mutant sites in the same cistron, no amount of recombination could make a virus with both cistrons functioning correctly. Benzer had therefore succeeded in dividing the *r*II gene into two clear-cut parts, each of which could function independently of the other.

The result of such work was to provide a very much more complicated picture of the gene. Very small regions, which can be defined through recombination or mutation experiments, are now respectively termed the recon and the muton. Much larger than these is the cistron region. Finally, there is the total region concerned with a given characteristic, such as the *r*II region. Which one should be called the gene? So far, the best answer seems to be to define the gene as whichever of these is appropriate for a given purpose.

But what are genes made of, and how do they work? To begin answering these questions has required the combined techniques of genetics and of molecular biology—which has stemmed from work in life chemistry that began early in the nineteenth century.

3

The Chemical Components

THE HUMAN BODY IS ABOUT 65 PER CENT WATER, 15 PER CENT proteins, 15 per cent fatty materials called lipids, 5 per cent inorganic materials, and less than 1 per cent carbohydrates. Proteins, the most complex of life substances, are found in, for instance, hair, horn, muscle, silk, and tendon. A familiar protein is white of egg, which is principally composed of ovalbumin. Lipids in the body include substances similar to butter or lard. Everyday instances of carbohydrates are potato starch and the sugar used to sweeten coffee.

During the early nineteenth century men were coming to recognize that these three classes of organic substances—lipids, carbohydrates, and proteins—occur in all living things, whether plant or animal. In 1838 Gerardus Johannes Mulder decided that protein was the most important of these and gave it its name, which means "primary" or "holding first place."

Today we know that he was right. Proteins are, indeed,

the group of substances that forms the basic machinery of life processes. They also confer individuality, for every animal or plant forms its own kinds of proteins. So, incredibly, does every individual: like fingerprints, no two individuals are identical in all of their proteins.

IS THERE A VITAL FORCE?

In their early exploration of the chemistry of living things, men were handicapped by their belief that substances of organic origin were mysteriously different from inorganic substances. They considered that all test-tube reactions must be different in kind from the chemistry of living things and that only a "vital force" could explain the creation of life substances. Thus the illustrious Swedish chemist Jöns Jakob Berzelius went on record as saying that life substances could never be made in the laboratory.

This belief in a vital force, known as "vitalism," was widespread among chemists in the early part of the nineteenth century. Maybe it was just a convenient way of sweeping under the carpet their bewilderment at the complexity of the chemistry of life substances. Unable to understand their chemistry, men "explained" their formation by the vital force—and so discounted the possibility that they would ever produce similar substances in the laboratory.

In this context, an obscure German research student named Friedrich Wöhler made a discovery which was to change men's outlook completely. While working in Berzelius' laboratory in 1824 he was investigating cyanic acid, which contains nitrogen. As part of this research he attempted to make the ammonium salt of the acid, ammonium cyanate, by the straightforward method of reacting aqueous ammonia with cyanic acid. To his amazement, he obtained oxalic acid and clear, colorless crystals which looked exactly like urea crystals—slender, four-sided, dull-pointed prisms. But how could this be? Urea, he knew, was a substance made and excreted by animals. Surely it was impossible for him to have made it in a test tube!

Maybe if Berzelius had not been one of the leading chemists of the day, and a firm supporter of vitalism, young Wöhler would have discussed his puzzling discovery with him. As it was, he preferred to wait until he could check and double-check his results.

Quitting Berzelius' laboratory in Sweden, Wöhler returned to his native Germany and became a teacher at the city trade school in Berlin. There he tested his findings and could discover no chemical difference between crystals of urea obtained from urine or those made by him from ammonia and cyanic acid.

He investigated other ways of making ammonium cyanate, using the well-tried method known to chemists as "double decomposition." For example, he reacted ammonium chloride with silver cyanate, hoping to obtain ammonium cyanate and silver chloride. He did the same with potassium cyanate and ammonium sulphate. In each case, however, he ended up with urea instead of ammonium cyanate. It was baffling.

Wöhler analyzed urea and found that it and ammonium cyanate contained the same chemical elements joined together in the same proportions. Each molecule consisted of four nitrogen atoms, two carbon atoms, eight hydrogen atoms, and two oxygen atoms. What, then, was the reason for the different chemical properties of the two compounds?

By coincidence, a similar conundrum had been posed earlier as a result of some other work of Wöhler's. In 1824 he had published an analysis of cyanic acid, shortly after another young German, Justus von Liebig, had worked out the composition of fulminic acid. It turned out that both acids contained one atom of carbon, one of oxygen, two of nitrogen, and four of hydrogen.

The great French chemist Joseph Gay-Lussac, in whose journal Liebig had published his results, noted the extraordinary similarity of chemical composition. He realized that cyanic and fulminic acids might be different even though they contained identical numbers of identical atoms if those atoms were differently arranged in the two substances. Liebig, who was working in Gay-Lussac's laboratory in Paris, checked Wöhler's work on cyanic acid and found that he got the same result. Evidently Gay-Lussac's hunch was correct! In celebration, the distinguished Frenchman and the young German researcher waltzed around the laboratory together.

Berzelius then checked the work on cyanic and fulminic acids himself. As a result, he accepted Gay-Lussac's explanation that the two substances must have the same atoms in different arrangements. He called such a phenomenon isomerism, a name which is still used for it.

In 1828 Wohler was finally sure of his work on urea, and wrote to Berzelius, in a sentence which has since become famous, "I must tell you that I can make urea without the use of kidneys, or of any animal, either man or dog."

Wöhler refused to speculate about why urea and ammonium cyanate, two such different chemicals, should have the same composition. Berzelius, however, immediately realized that this must be another example of isomerism.

> This is a very remarkable transformation [Berzelius wrote]. We are here dealing with a substance having exactly the same composition as a salt. Yet it is no salt, for it is not decomposed by bases, acids, or salts, as it should be if it were a salt. On the contrary, it has all the properties of an organic molecule and combines partly with acids and partly with bases. This is an example, therefore,

that shows how compounds containing the same relative number of atoms of the elements can differ in their chemical properties, because of the different relation of the atoms to one another in a molecule.

Berzelius also recognized that his former student had made a breakthrough in the chemistry of life substances. Thus in 1830, when Wöhler had just missed discovering a newly identified metal, vanadium, the Swedish chemist wrote consolingly: "The chemist who has discovered a way to produce an organic body artificially can well afford to forego all claims to the discovery of a new metal."

Other chemists also accepted the significance of Wöhler's discovery. Although the theory of vitalism was still to have currency for many years, its downfall had begun.

Liebig and Wöhler now joined forces to explore the chemistry of still other organic substances. They were so successful that in 1838 they were moved to make a prophecy which, though astonishingly bold at the time, is one we now know to be true:

> The philosophy of chemistry [they said] must draw the conclusion that the synthesis of all organic compounds must be looked upon not merely as probable, but as certain of ultimate achievement. Sugar, salicin, morphine will be artificially prepared. As yet we are ignorant of the road by which this end is to be reached, since the constituents required for building up these substances are not yet known to us; but these the progress of science cannot fail to reveal.

Liebig also took a swipe at the still-existent theory of vitalism. He affirmed that living things operated according to chemical laws and if those could explain how organic substances were formed, then he saw no need to invoke any mysterious vital force.

> Vitality [wrote Liebig] is the power which each organ possesses of constantly reproducing itself; for this it requires a supply of

substances which contain the constituent elements of its own substance, and are capable of undergoing transformation. All the organs together cannot generate a single element, carbon, nitrogen, or a metallic oxide.

To understand life, insisted Liebig, it was first necessary to understand the power and subtlety of chemical action. The molecular biologist of today would agree.

LIPIDS

Although chemists agreed on the prime importance of proteins, these substances are so complex that at first they could make little headway in elucidating their structure. The chemists' earliest successes were not with proteins but with lipids. These contain carbon, hydrogen, and oxygen only.

Michel Eugène Chevreul, a Frenchman who lived to the ripe old age of a hundred and three, began work on their structure as early as 1813. He proved that they contain a mixture of substances called triglycerides. These are compounds of glycerol (familiar to us as glycerine) and various acids having a long carbon chain as their backbone and an acid group at one end—the so-called fatty acids.

A familiar example of a fatty acid is butyric acid, which occurs in butter. Its formula can be written

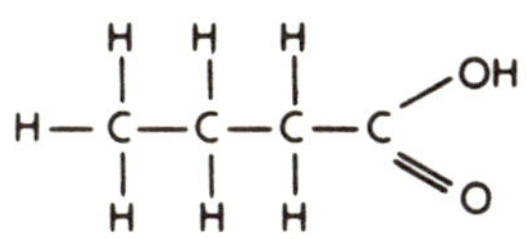

where H stands for a hydrogen atom, C for a carbon atom, and O for an oxygen atom. Each line represents a chemical bond. The acidic part of the molecule is the –COOH group.

Glycerol is a member of the class of substances called alcohols. One of the simplest alcohols from a chemical point of view is ethyl alcohol, available in diluted form at the local liquor store. The formula of glycerol can be set out as

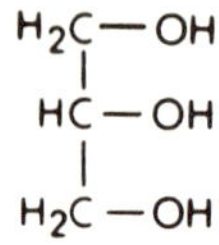

To form a triglyceride glycerol links up with three fatty acid molecules. This link-up is made by the elimination at each bond of a molecule of water, H_2O–a process known as condensation. The reaction can be portrayed as follows, using R to stand for the long chain of some fatty acid:

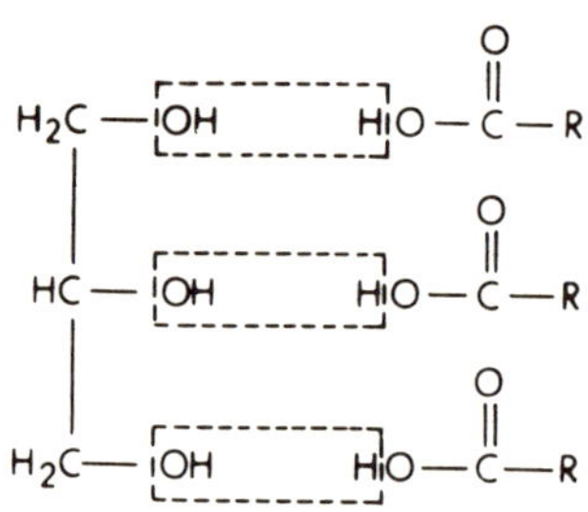

In the 1850s another French scientist, Pierre Eugène Marcelin Bertholet, helped to sound the death knell of vitalism by synthesizing lipids in his laboratory.

Using glycerol as a starter, he joined it to various acids which do not occur in natural fats–and made new lipids. In other words, he not only copied nature but went beyond it. (Much of the chemical industry today is based on just that ambition.)

CARBOHYDRATES

Like lipids, the carbohydrates contain only carbon, hydrogen, and oxygen. However, their molecular structure remained a mystery until the 1870s, when the great German researcher Emil Fischer, one of the outstanding chemists of all time, turned his attention to them.

Fischer was a man of contagious enthusiasm and everyone who worked with him was affected by him. He was a large man physically, and mentally too, being impatient of trivialities. If any one man carried out Liebig's precept that to understand life we must first understand its chemical subtleties, that man was Emil Fischer.

Only one year after graduation from the University of Strasbourg in 1874, Fischer discovered phenylhydrazine. He found that this chemical combined with sugars, which in living things are usually associated with other materials. By reacting sugars with phenylhydrazine, extracting the sugar plus phenylhydrazine, and then splitting off the phenylhydrazine, Fischer isolated various sugars. This was the essential first step toward studying them intensively.

In this way Fischer separated glucose, or grape sugar; maltose, or malt sugar; fructose, or fruit sugar; and many others. All of these, he found, had the same chemical composition: six carbon atoms, six oxygen atoms, and twelve hydrogen atoms. Yet their chemical properties were different. Here, evidently, was another case of isomerism!

How many isomers of these six-carbon sugars could there be? Assuming a backbone of carbon atoms, Fischer figured out that there could be a total of sixteen. He determined to synthesize them.

Adolf von Baeyer, Fischer's former professor, suggested that in nature sugars might be synthesized from formaldehyde. Using this idea as a starting point, Fischer succeeded in synthesizing fructose and glucose as early as 1887. Eventually he synthesized twelve out of the possible sixteen isomers—an amazing feat. Also, his work involved great insight into the importance of molecular structure.

Two of the isomers, glucose and fructose, can be represented as follows:

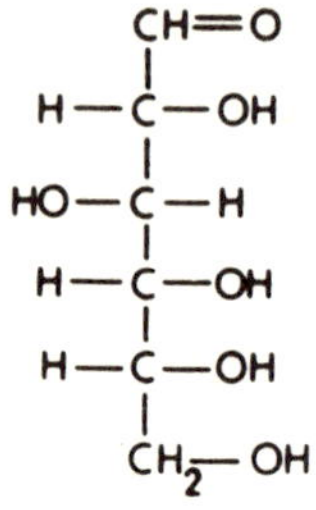

glucose

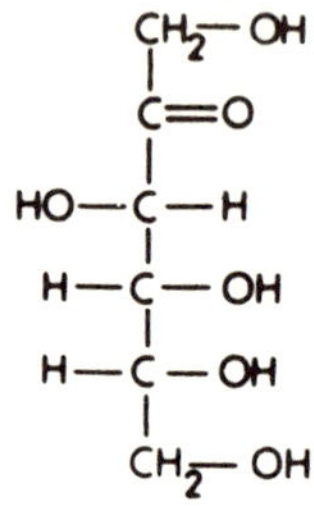

fructose

These are not very accurate representations, however. Even in 1883, Bernard Tollens suggested that glucose is not really a straight chain of carbon atoms with side arms, but a ring structure. In the 1920s and 1930s Walter Norman Haworth, a British scientist, worked out that the carbon atoms and one oxygen atom form a flat hexagonal ring, from which the side groups stand out at right angles above and below:

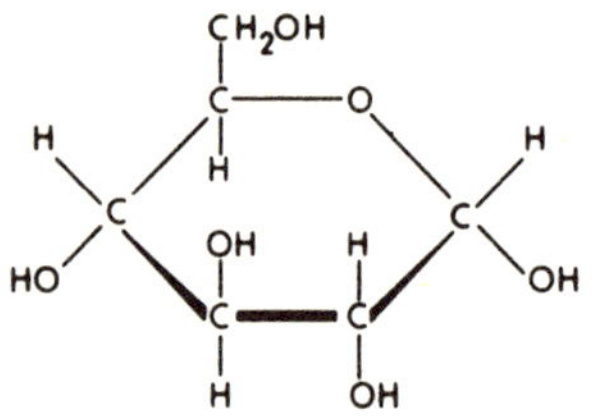

glucose

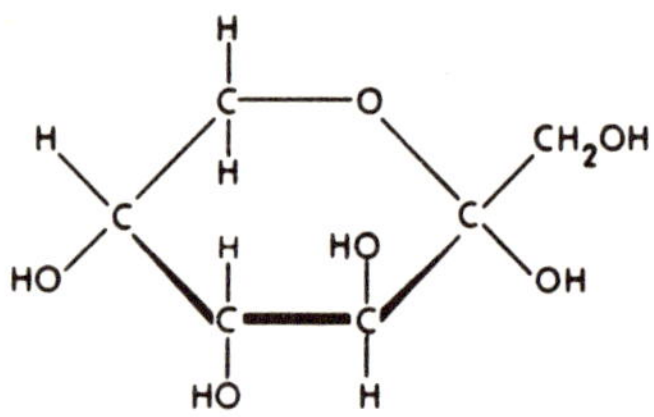

fructose

The thick lines indicate that the hexagonal ring is standing out at right angles to the plane of the paper.

This three-dimensional picture helps us to visualize a very important chemical property of the six-carbon sugars: they can join together by condensation to make larger units. This is shown on the next page, with the carbon atoms in the ring omitted for simplicity.

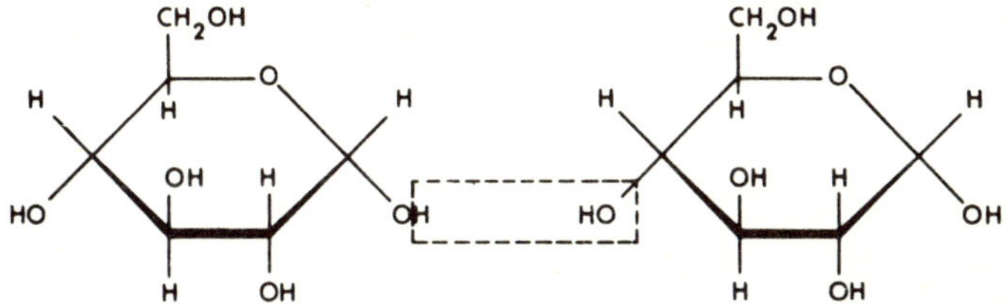

In this way glucose and fructose join to form sucrose (cane sugar), used to sweeten our coffee. A two-sugar molecule of this type is called a disaccharide. Strings of three sugars, or trisaccharides, are common. Chains of many sugars are polysaccharides, familiar examples being potato starch and cotton (cellulose).

PROTEINS

The third great category of life substances, the proteins, not only contain carbon, hydrogen, and oxygen—as do the lipids and carbohydrates—but nitrogen and sulfur. When Fischer began work on proteins, it was already known that they too are built up from simpler units, rather like the polysaccharides, but the units are amino acids.

A French chemist, M. H. Braconnet, was the first man to isolate an amino acid. Back in 1819, he had boiled cellulose with acid and obtained glucose. From this he realized that cellulose must be formed from glucose units, although he could not tell how they were linked. He then went on to treat a protein, gelatin, in the same way and obtained a sweet, crystalline substance. This could not be a sugar because he obtained ammonia (NH_3) from it, thus proving that it contained nitrogen. He called the substance glycine. Soon after, he obtained another crystalline substance by heating muscle tissue with acid: he called this leucine.

Many other amino acids were isolated and chemically examined during the nineteenth century. In all instances they were obtained by what chemists term hydrolysis, meaning splitting with water. Water is in fact used up during the reaction, which is the opposite of condensation. Acids and alkalies can be used to bring hydrolysis about, and so can prolonged boiling with water.

All of these amino acids were found to contain one or more acid carboxyl (–COOH) groups and one or more amino ($-NH_2$) groups, which are basic as opposed to acidic. Amino acids therefore have a dual quality. Fischer showed that this is the secret of their linkage. The basic amino group of one amino acid joins through condensation with the acidic carboxyl group of another.

Thus glycine condenses with glycine to form a two-unit chain called glycylglycine:

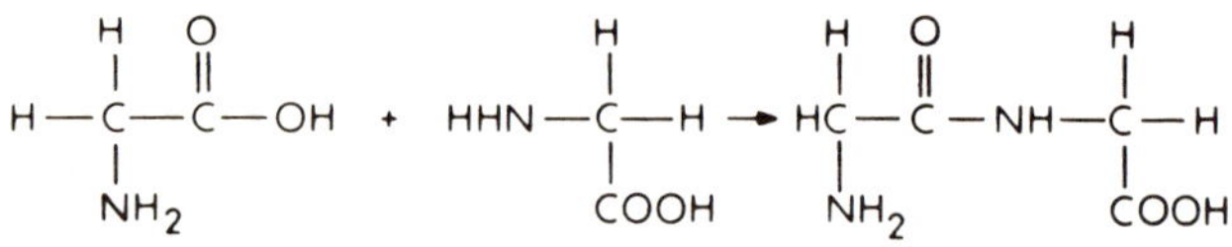

glycine glycine glycylglycine

This link between a carboxyl group and an amino group is called a peptide bond. Chains of amino acids joined in this way are called polypeptides. By repeating the condensation process, Fischer built up longer and longer polypeptides until, in 1907, he succeeded in making a chain of fifteen glycine units, or residues as they are called, plus three leucine residues.

Yet even this remarkable achievement was not a protein. While having protein-like properties, it was evidently much smaller. It was, in fact, similar to the fractions of natural proteins found after protein has been partially digested in the stomach. Nevertheless, Fischer's accomplishment was a tremendous one for the time in which he worked.

This was as far as Fischer could go. Today we understand the awesome complexity of the task he had begun. There are twenty common amino acids, and the number of combinations they can theoretically make is truly astronomical. This is why protein is so suitable for making living tissues: there is virtually no limit to the variety obtainable with it. The size of a protein molecule, as indicated by its molecular weight, makes it a giant, too. A medium-sized protein molecule has about 5,000 atoms arranged in some 400 amino acid residues.

Because of their highly complicated structure, proteins were long regarded by biochemists as too difficult to analyze in detail. Then, in the 1940s, Albert Charles Chibnall and his colleagues at the University of Cambridge in England attempted to discover their structure. One member of the group, twenty-six-year-old Frederick Sanger, set himself the task of working out the sequence of residues in insulin. This is a relatively small protein, containing only fifty-one residues. After ten years he succeeded and was awarded a Nobel prize for his work. The sequences for more than ten proteins are now known.

NUCLEIC ACID

Proteins, lipids, and carbohydrates–for many years chemists and nutritionists alike regarded these as the big three among life substances. Yet all the while a fourth material which ultimately was to prove more significant than any of these was stored, almost unnoticed, on the laboratory shelves.

It was in 1869 that a young Swiss biochemist named Friedrich Miescher discovered this substance, which he christened nuclein because it seemed to exist only in the nuclei of cells.

At this time Miescher was making a systematic analysis of cell nuclei, using as his source material pus from surgical

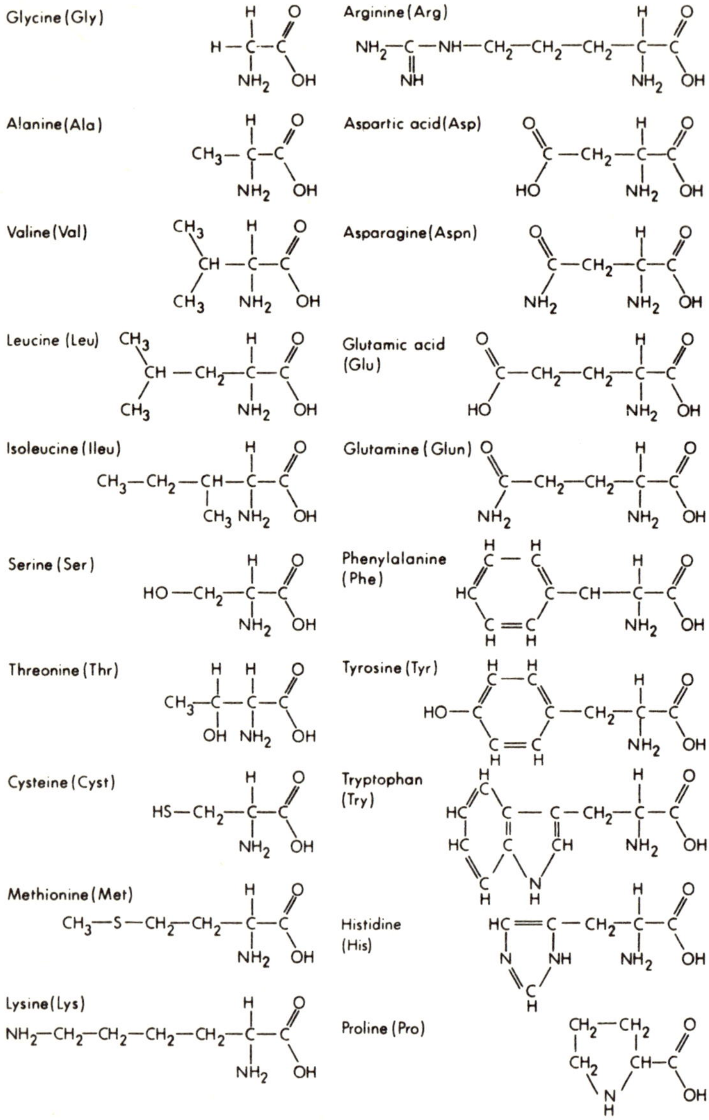

Figure 8. The names and formulas for the twenty commonly occurring amino acids, together with abbreviations for the names.

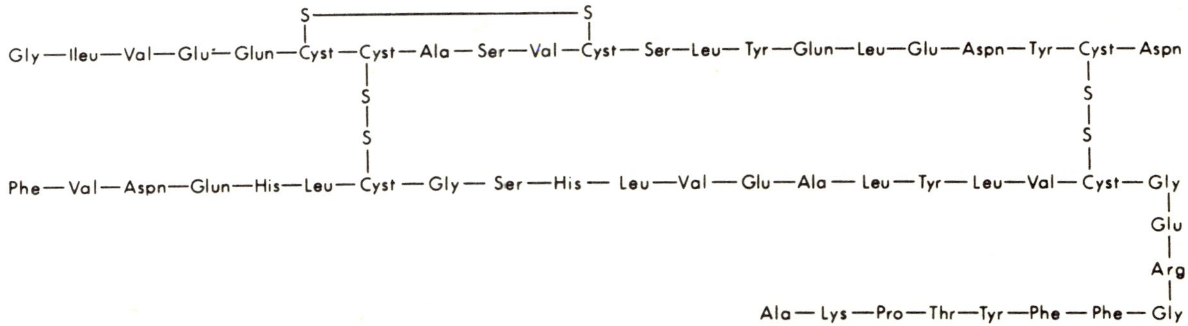

Figure 9. Sequence of amino acids in the insulin molecule of the ox. Sulfur bridges (S – S) link parts of the molecule. The shape of the molecule is not indicated.

bandages discarded after use in the local hospital. Obtaining this pus was not a job anyone would envy, and Miescher himself confessed, "I feel as if I am mired in a swamp." However, he persevered with his bizarre supply of nuclear material and thus was led to discover a fourth fundamental category of life substance.

To separate the nuclei from a suspension of the pus cells, Miescher used to digest the protein-rich remainder of the cell body away from the nucleus by means of pepsin, a protein-splitting component of the stomach juices. To his surprise, he found that the nucleus contained a material which was not fat, carbohydrate, or protein. It was unlike any other known life substance in that it contained phosphorus. In solution it formed an acid. For this reason, its name was later changed from nuclein to nucleic acid.

Miescher needed a plentiful supply of nuclear material before he could carry out further research on nucleic acid. He found this supply in salmon sperm. Each spring he visited the Rhine falls above his home town of Basel at the time when the salmon were jumping the falls. Local fishermen were there to catch the salmon and Miescher obtained with their help a supply of salmon sperm which lasted him through the following year.

The sperm was a rich source of nucleic acid because each sperm is virtually all nucleus. To prepare the nucleic acid, Miescher dissolved the sperm heads in strong salt solution and then precipitated fibers of nucleic acid by adding water. He had to keep his preparations as cool as possible and so, in those pre-refrigerator days, he worked through the winter in an unheated room. In a letter to a friend he described his methods:

> When nucleic acid is to be prepared, I go at five o'clock in the morning to the laboratory. . . . No solution can stand for more than five minutes, no precipitate more than one hour being placed under absolute alcohol. Often it goes on until late in the night.

Only in this way do I finally get products of constant phosphorus proportion.

Working in these rigorous conditions, Miescher succeeded in analyzing the chemical elements present in nucleic acid. He found that, besides phosphorus, there were carbon, oxygen, hydrogen, and nitrogen. This was as far as he could go.

A German chemist named Albrecht Kossel continued research on nucleic acid. He discovered that it contained combinations of atoms built around two basic carbon-nitrogen configurations known as the purine and pyrimidine groups.

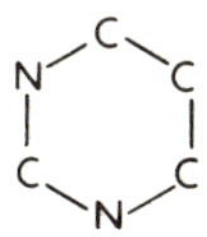

pyrimidine skeleton

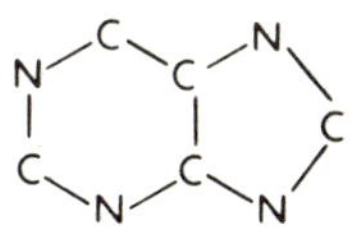

purine skeleton

The chemistry of the purine group had already been worked out quite independently by Emil Fischer. He had been led to this from a study of the chemistry of uric acid. He found that this substance, with the formula $C_5H_4N_4O_3$, had derivatives in which the number of oxygen atoms was steadily reduced. These derivatives were xanthine, $C_5H_4N_4O_2$, and hypoxanthine, $C_5H_4N_4O$. From this, Fischer reasoned that there should be a parent group embodying the basic formula $C_5H_4N_4$ and having no oxygen at all. He christened this purine in 1884 and four years later succeeded in synthesizing it. Purines occur throughout nature, and it was therefore no great surprise when Kossel found them in nucleic acid.

Kossel isolated two pyrimidines, which he named cytosine and thymine, and two purines, which he named adenine and guanine. For this work he received the Nobel Prize in Medicine and Physiology for 1910. Another pyrimidine, uracil, was later discovered in some forms of nucleic acid. A curious fact about all of these was that, despite their being present in an acid substance, they themselves were basic in chemical character—that is, the opposite of acidic.

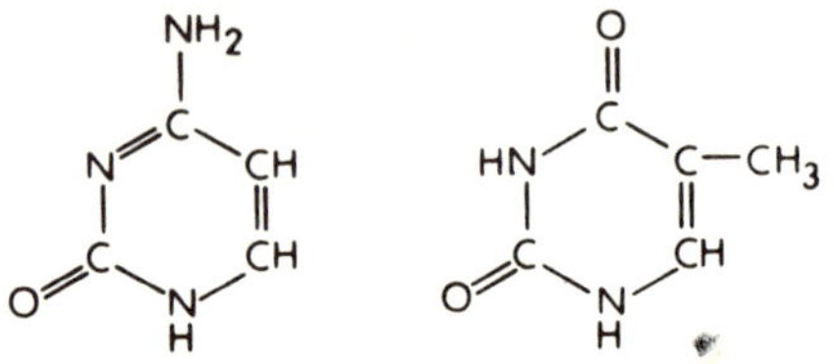

cytosine thymine uracil

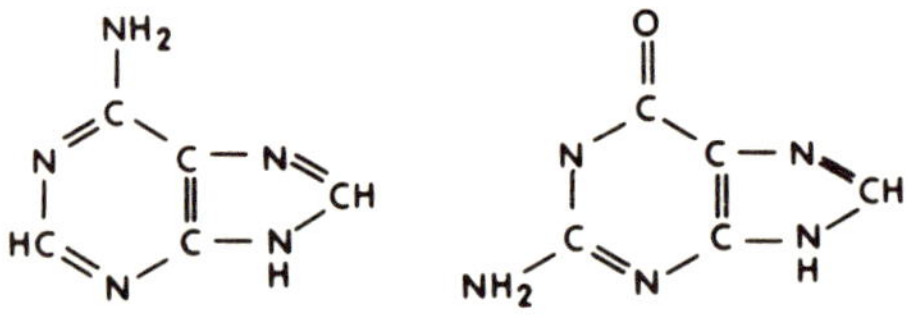

adenine guanine

The thread of research was now taken up by Phoebus Aaron Theodore Levene, a Russian-born biochemist working at the Rockefeller Institute in New York. In 1910 he identified a five-carbon sugar, ribose, in nucleic acid. This, too, was no stranger to chemists as the indefatigable Emil Fischer had synthesized it in 1901.

Later, Levene discovered that not all kinds of nucleic acid contain ribose sugar. Some have instead a related sugar which has one less oxygen atom and is therefore called deoxyribose. Like ribose, this too had been synthesized by Fischer but had hitherto remained a laboratory curiosity.

Thus it was now clear that there were two kinds of nucleic acid, one containing ribose sugar and so called ribose

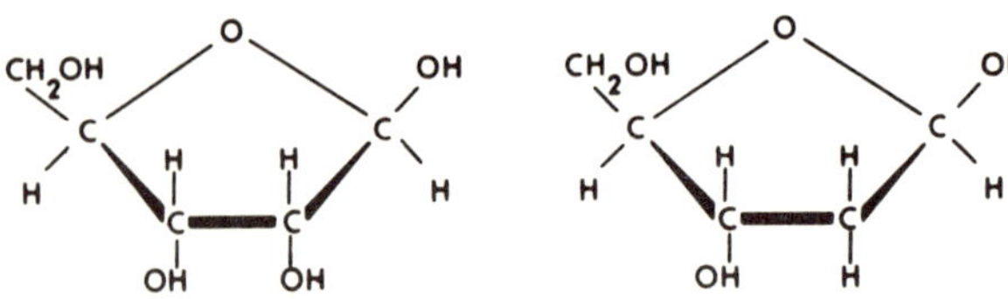

ribose deoxyribose

nucleic acid, or RNA, and the other containing deoxyribose sugar and so called deoxyribose nucleic acid, or DNA.

The purines, pyrimidines, and five-carbon sugars in these nucleic acids accounted for the presence of carbon, hydrogen, oxygen, and nitrogen, but not for the phosphorus which Miescher had also identified. What part could the phosphorus play in the molecule?

Levene came up with a suggestion about this. He figured that, from his general knowledge of chemistry, the phosphorus would probably be present in a phosphate group of phosphorus, oxygen, and hydrogen. He further surmised that the sugar could be a bridge linking the purine (or pyrimidine) and the phosphate group. The resulting structure was called a nucleotide.

Levene also suggested how the various nucleotides might join together. The key, he said, was the phosphate group, which could link one base-sugar combination to the next. In RNA, the base could be adenine, guanine, cytosine, or uracil.

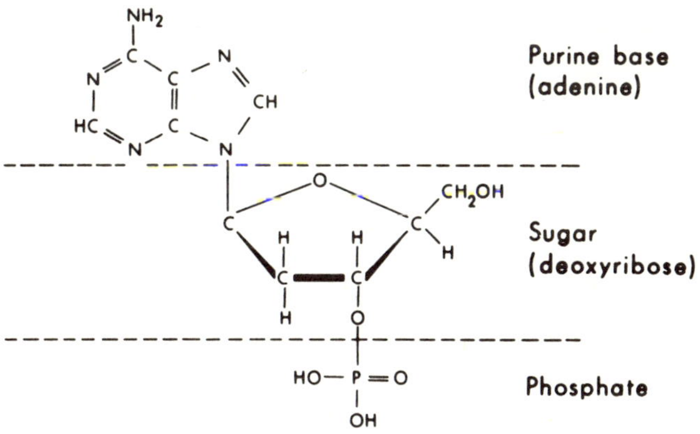

Figure 10. Structure of deoxyadenylic acid, the adenine-containing nucleotide in DNA. H = hydrogen, C = carbon, N = nitrogen, O = oxygen, and P = phosphorus.

In DNA, the base could be adenine, guanine, cytosine, or thymine.

Thus Levene envisaged four different nucleotides for RNA–adenylic acid (containing adenine), guanylic acid (containing guanine), cytidylic acid (containing cytosine), and uridylic acid (containing uracil). In the same way, there were four more nucleotides for DNA–deoxyadenylic acid, deoxyguanylic acid, deoxycytidylic acid, and deoxythymidylic acid (containing thymine).

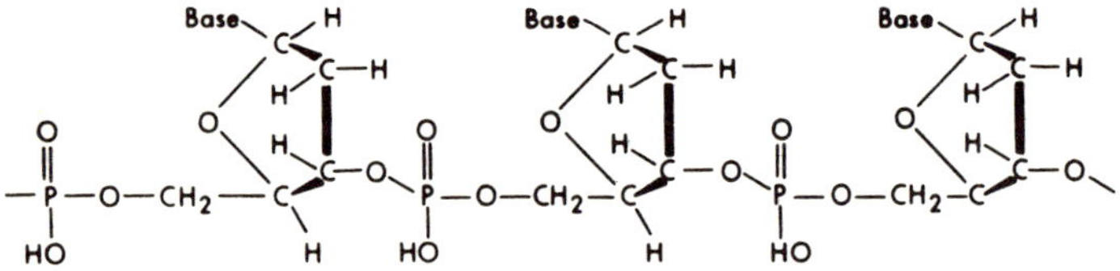

Figure 11. How a chain of nucleotides is formed by linking between phosphate groups.

Because of the way these suggestions of Levene's fitted in with general chemical knowledge, they were fairly well accepted by chemists. However, they were not fully verified until the early 1950s when a British chemist, Alexander Robertus Todd, synthesized structures that obeyed Levene's formulas exactly. He found that they were identical with the material obtained from nucleic acid. For this work Todd was awarded the 1957 Nobel Prize in Chemistry.

For many years before that, most scientists had paid little attention to nucleic acid. They preferred to concentrate their attention upon proteins, for these were widely regarded as the key to understanding the nature of life. The main reason for their interest lay in the marvelous qualities of those particular proteins called enzymes, which are certainly an essential mechanism in the functioning of the living cell.

4

The Role of Enzymes

An outstanding characteristic of living things is that they survive by taking in substances unlike themselves and using them as food. As Walter de la Mare rightly observed:

It is a very odd thing–
As odd as can be–
That whatever Miss T. eats
Turns into Miss T.

Green plants take in carbon dioxide from the air, and water, nitrogen, phosphorus, and other inorganic elements from the soil. With the energy of sunlight, they synthesize from these simple substances all the complicated chemical compounds, including protein, which they require.

Animals take in the complex fats, carbohydrates, and proteins already built up by plants or other animals. From these they make their own versions of these substances and obtain the energy they need.

Living things are, in fact, chemical factories. Further, they can execute a multitude of chemical changes with an efficiency and speed that cannot be rivaled by the chemist in his laboratory. Using powerful chemicals, high temperatures, and maybe electricity, the chemist frequently will take months to synthesize an organic compound. By contrast, a single bacterium can make a copy of itself, with all the chemical synthesis that entails, in just half an hour.

As an example of the chemical wizardry constantly going on inside ourselves, consider phenylalanine, an amino acid which we have to have in our food because we cannot synthesize it. Phenylalanine is oxidized in the liver to tyrosine. (If it is not, idiocy results.) The thyroid gland converts this tyrosine to thyroxine, which is essential for control of the body's rate of working, or metabolism. The adrenal glands, on the other hand, use tyrosine to make adrenaline, which governs blood pressure, sweating, and general physical readiness. The skin uses tyrosine another way: it produces melanin, the pigment giving brown skin color.

Chemical reactions in living things are carried out in closely controlled sequences. Each step in any sequence is governed by the presence of a particular chemical catalyst which brings about the reaction without itself being changed. In living things, such catalysts are called enzymes.

DISCOVERY OF ENZYMES

One of the earliest examples of enzyme action to be discovered —though not then recognized as such—was that of pepsin. This protein-splitting substance is present in stomach juices.

In the latter part of the eighteenth century, the great Italian physiologist Spallanzani—the man who fitted silk pants on mating frogs and toads—tried feeding hawks with meat enclosed in little wire cages. After the birds had had time to digest the meat, he made them regurgitate the cages and found that the meat had disappeared. Since it could not have been ground up by the birds, it had evidently been liquefied by some agent in their stomach juices. In 1836 later workers identified pepsin as the agent concerned.

By this time other substances had been found which instigated chemical reactions in living things. The first to be given a name was diastase, which hydrolyzes starch into sugar. This was found in 1830, when Anselme Payen and J. F. Persoz used alcohol to precipitate this substance from malt extract. Then Liebig and Wöhler found emulsin, a substance present in almonds which breaks down a crystalline compound in them called amygdalin, forming almond oil and prussic acid. Ptyalin, which is present in our saliva and breaks down starch, was described in 1831.

A characteristic of all these newly discovered substances was that very small amounts of them would produce very large effects. For example, a solution containing only one ten-millionth of an ounce of pepsin clots milk. Rennin, which is present in calves' stomachs and is used for making junkets and cheese, can clot ten million times its own weight of milk in five minutes. Even more spectacular a performer is catalase, one molecule of which can decompose over two million molecules of pure hydrogen peroxide every minute.

Berzelius, the great Swedish chemist, pointed out that many chemical reactions could proceed to a noticeable extent only if there were present a relatively small amount of some other chemical which was not itself affected chemically by the reaction. For example, hydrogen and oxygen gas will combine in the presence of platinum metal to form water.

In 1835 Berzelius drew up a whole list of such chemical reactions, both organic and inorganic, and suggested that they seemed to operate by means of a new force. He wrote:

This new force is common to organic and inorganic nature. I do not know that it is a force entirely independent of electrochemical affinities of matter; I believe, on the contrary, that it is only a new manifestation, but as long as we cannot see their connection and mutual dependence, it will be more useful to designate it by a separate name. I will then call this force "catalytic force."

Berzelius specified that the agents of this force, the catalysts, somehow excited the substances which were chemically changed in the reaction and brought out qualities which had lain dormant in them. How they could do this, he did not guess.

One organic catalyst which caught his attention was diastase. He was struck by the fact that this hydrolyzes starch more efficiently than sulfuric acid. With amazing insight, Berzelius suggested that scientists would probably eventually find that all the substances in living organs are produced by the action of such powerful organic catalysts. Today, we know that he was right.

ARE ENZYMES ALIVE?

The old controversy of vitalism, increasingly discredited since Wöhler's synthesis of urea, was given new life because of these discoveries of organic catalysts.

In 1839 Liebig took hold of Berzelius' new definition of a catalyst and linked it with his own research on emulsin. He suggested that fermentation, decay, and putrefaction occurred because of organic catalysts. Then he went on to speculate that these catalysts were energy-rich substances which somehow transferred their energy to the substances they affected, so causing them to split up.

Drawing analogies from inorganic chemistry, Liebig compared the action of organic catalysts to that of heat and acids.

He also pointed out that some catalysts affected only specific substances, and he suggested that there might be structural similarities between catalyst and substrate—the substance acted upon—which made each catalyst specific in its action. (This idea was later adopted by others.)

In all of this, Liebig was at pains to stress that the functioning of organic catalysts should be explicable on strictly mechanistic lines. In his view there was no need to invoke some vital force. He went even further and denied that living organisms were to be found in fermenting mixtures—fermentation being a prime example of organic catalysis. When taxed with the presence of yeast cells, which ferment sugar to form alcohol, he denied for a long time that these cells were living. Even when persuaded of the contrary, he refused to agree that fermentation occurred while they were alive. It only occurred, he said, when the cells were dead!

This was plain stubbornness and Liebig was able to hold out so long only because of the inadequacy of his microscopic investigations. The great French chemist and founder of bacteriology, Louis Pasteur, was a master of the microscope, and he disagreed profoundly with Liebig. Pasteur had been studying the way in which milk goes sour by fermentation, with the formation of lactic acid. In August 1857, he presented a short paper before the faculty of science at the University of Lille which is a classic in the history of bacteriology. Three months later he repeated it before the Académie des Sciences in Paris.

In this paper Pasteur showed that the lactic acid was produced by a living ferment. This ferment, he explained, would grow in a clear nutrient broth where it multiplied to give a population of microscopic living things—bacteria. Once grown in adequate amount, this living mass would produce lactic acid at an astonishing rate. He wrote:

The purity of a ferment, its homogeneity, its free unrestrained development by the aid of food substances well adapted to its individual nature, these are some of the conditions which are essential for good fermentation.

Pasteur also pointed out that if the living cells he had observed were destroyed by heat or removed by filtration, the fermentation process ceased. Without life, he concluded, there was no fermentation. And what was true for the fermentation of milk by lactic acid bacteria was also true, in his view, for the fermentation of sugar solution by yeast.

These findings were of course in complete contradiction to Liebig's views. One of Pasteur's trump cards in the ensuing controversy between the two men was that fermentation produces optically active compounds. This means that the compounds will bend polarized light to one side as this light passes through a crystal, or a solution, of the compound.

As a young graduate of the Ecole Normale in Paris, Pasteur had discovered that some chemical substances can have identical chemical properties and yet differ in the direction in which they bend polarized light. He prepared crystals of tartaric salts which were mirror images of each other in shape and bent polarized light in opposite directions—one to the left, the other to the right. And if the crystals were mirror images of each other, so—reasoned Pasteur—must be the individual molecules composing them. He had discovered a new form of isomerism—not chemical, like that of urea and ammonium cyanate, but optical and structural.

In 1857 Pasteur chanced upon a highly significant fact. A solution of an organic salt, calcium tartrate, he had in his laboratory was spoiled by a mold growing upon it. Most researchers would have thrown the liquid away, but Pasteur wondered whether the mold would ferment both the right- and left-handed forms of the calcium tartrate. He found, to his intense interest, that only the right-hand form was used by the mold. He immediately grasped the significance of this:

> Here then we see molecular asymmetry, a property peculiar to organic matter, influencing a physiological process . . . thus, the concept of the influence of molecular asymmetry of natural organic products is introduced into physiological studies, through this important criterion [optical activity], which forms perhaps the only sharply defined boundary which can be drawn at the present time between the chemistry of dead and that of living matter.

Pasteur had realized that the selective usage of the tartrate by the mold meant that the mold was, as it were, right-handed. In a laboratory, any compound that can exist in both right- and left-hand forms is always synthesized artificially in equal amounts of each form. This means that if a material occurs in only one of its two optically active forms it must have been synthesized by a living organism. Here, then, is a subtle test for the influence of a living thing.

Pasteur applied this test to the materials made during fermentation by yeast and by lactic acid bacteria. He found that when two isomeric forms could exist, only one in fact was produced. He therefore concluded that fermentation was a living process.

Since scientists knew that powerful organic catalysts existed in, for instance, the stomach juices, it now seemed that such catalysts were different in kind from the ferments described by Pasteur. The stomach juices, after all, could not be thought of as alive, even though they were produced by a living creature. Scientists therefore came to distinguish between organic catalysts which could act independently of living cells, which they termed "unorganized ferments," and others which they believed were the living cells themselves and which they called "organized ferments."

In 1878 Wilhelm Kühne christened the "unorganized ferments" enzymes, meaning "in yeast," for by then yeast had been found to contain a rich variety of organic catalysts which could act without the living yeast.

The controversy concerning "unorganized ferments" and "organized ferments" remained unresolved. In 1888 a German named Eduard Buchner put his mind to the problem and determined to try to extract from yeast the active principle which he believed must cause fermentation. Adolf von Baeyer, in whose laboratory he was then working, pointed out discouragingly that scientists had already tried to do just this for forty years and had failed. However, Buchner refused to quit.

In the fall of 1896 he spent a vacation with his elder brother Hans in the Buchners' home town of Munich. His brother was director of the Hygiene Institute and Buchner seized the opportunity of using the Institute's laboratory resources to experiment further with yeast. By bitter experience he already knew that there were many snags: too high a temperature, or chemically active solvents, would destroy any active principle in the yeast cell. Therefore he had to find another method of attack.

He decided to try grinding the cells and squeezing out the juice. As an abrasive, he used quartz sand and kieselguhr, a fine clay, grinding these up with the mass of yeast in a mortar. Then he wrapped the resulting doughy mass in a cloth and put it under a pressure of 1,250 pounds per square inch in a hydraulic press. A juice seeped out, and within a few hours Buchner had more than half the cell content in liquid form.

Now he added sugar to the liquid, to test for fermentation. The result was obvious in less than a quarter of an hour: bubbles of carbon dioxide were forming in the mixture. He had extracted the agent of fermentation from the living cell and proved that Liebig had been right and Pasteur wrong. His extract had only one fortieth of the power of the original yeast—but it worked!

Buchner portrayed his findings in bold colors. "The cells

of plants and animals," he wrote, "appear with increasing distinctness as factories where in separate workshops all kinds of products are produced. The foremen in this work are the enzymes."

In 1907 Buchner was awarded a Nobel prize for this work, described by the committee conferring the award as "his chemico-biological investigations and his discovery of cell-less fermentation."

ENZYMES ARE PROTEINS

In spite of all this, chemists remained reluctant to think of enzymes as being like other chemicals. The theory of vitalism might be exploded, fermentation might be only a nonliving process, but enzymes still seemed special. Their ability to catalyze reactions was uncanny, almost magical.

This was the atmosphere in the 1920s when James Batchellor Sumner at Cornell University decided to try isolating an enzyme pure. He chose urease, which is present in jack beans. (In accordance with a naming convention proposed back in 1883 by a Frenchman, Pierre Emile Duclaux, an enzyme is commonly designated by the name of the substrate it acts on, plus the ending "ase." Urease acts upon urea—in fact, it breaks it down into ammonia and carbon dioxide.)

Sumner ground up his beans in an old-fashioned coffee mill, added water, and filtered out the solid particles. Then he used acetone to dissolve the enzyme left in the liquid, extracted the acetone, and found tiny eight-sided crystals appearing in it. These were pure urease. Excitedly, Sumner rushed to the telephone and told his wife that he had succeeded in crystallizing an enzyme for the first time.

The crystals he obtained were pure protein. There was nothing mysterious, nothing vitalistic, about them. Yet even in science old ideas die hard. The accepted leading authority of the day on enzymes, a German named Richard Willstätter, would not accept Sumner's findings. He and his students summarily dismissed the discovery, saying that the crystals were not of the pure enzyme but of a protein which somehow carried the enzyme. As a guest lecturer at Cornell, Willstätter publicly rebuffed Sumner.

In answer, Sumner wrote an article justifying his interpretation of his work, had a friend translate it, and sent it to the German scientific journal *Die Naturwissenschaften* for publication. In spite of this, another visiting lecturer from Germany, Hans Pringsheim, repeated Willstätter's performance at Cornell. This time, however, Sumner asked permission to address Pringsheim's class and put his findings before them. He was able to convince Pringsheim and his class that he was correct.

Even so, some biochemists remained skeptical. More than ten years later, they were still refusing to admit that enzymes were just proteins, no more.

The first man to find out that other enzymes are also proteins was John Howard Northrop of the Rockefeller Institute for Medical Research. In 1930 Northrop succeeded in purifying pepsin and found that it, too, was a protein. More enzymes were purified after this—and all of them turned out to be protein.

By 1947 the epoch-making nature of Sumner's work was fully accepted and he and Northrop shared the Nobel Prize for Chemistry that year.

CHEMICAL STRUCTURE OF ENZYMES

The next goal was to elucidate the exact chemical structure of an enzyme. This involved finding out the amino acids used to build it up, and the order in which they were joined in the enzyme molecule.

To analyze the amino acid sequence in any protein, techniques are used which were pioneered by Vernon Martin Ingram at the University of Cambridge in England when he worked out the chemical structure of hemoglobin, the red oxygen-carrying protein in the blood. These techniques involve using enzymes to break the protein molecule into smaller fragments.

The great advantage of these enzymes is that they break linkages between only certain kinds of amino acids and so allow us to control the way in which the peptide chains are fragmented. From this it follows that it is easier for us to deduce how these chains are joined together in the complete molecule. The smaller fragments are an advantage because they can be chemically analyzed more easily than large molecules.

William Howard Stein and Stanford Moore of the Rockefeller Institute have used this method to work out the complete amino acid sequence in ribonuclease, an enzyme that attacks RNA.

To help them, the Rockefeller researchers built an extremely ingenious apparatus in which the mixture of amino acids and peptides left after the protein-splitting enzymes had been at work was washed through a tall column packed with potato starch. As each amino acid and peptide had its own affinity for the starch surface, these substances were held back in the starch column by varying amounts. Differing quantities of solution were needed to wash each substance through the column and so they appeared neatly separated from each other in sequence at the bottom. Once separated, each could be chemically identified.

To determine the order in which all of these fragments were originally assembled, the researchers had to build up a picture from many overlapping pieces. Their reasoning was similar to that which we would use if we had to identify a word of twelve letters and knew that it contained the se quences RIBONU, NUCL, and CLEASE. Clearly, the word must be RIBONUCLEASE.

The Rockefeller Institute workers also determined that there were cross-links between parts of the amino acid chain. These links were of a kind well known to exist between sulfur atoms in individual amino acids. Their presence caused the chain to double up into convolutions.

SHAPE OF ENZYMES

That the shape of an enzyme molecule might be important was first suggested by Emil Fischer. By 1887 he had succeeded in synthesizing fructose and glucose in his laboratory, and he then went on to make other six-carbon sugars. In 1890 he followed up Pasteur's observations on the asymmetry of nature by growing yeast in a mixture of right- and left-handed forms of two sugars, mannose and fructose. He found that the yeast used up only the right-handed form of each sugar.

Evidently, Fischer reasoned, the shape of the sugar molecule mattered to the yeast. But why should it? The reason must be that the catalyst molecules inside the yeast would work only with sugar molecules of a particular shape. There must be some mechanism between catalyst and substrate akin to that between a lock and a key. (This was an extension of a guess Liebig had previously made.)

Having framed this lock-and-key concept with sugars, Fischer went on—brilliantly—to use it with polypeptides. He synthesized these out of natural and artificial amino acids. Then he tested their similarity to real protein fragments by studying how they were affected by enzymes which he extracted from animal tissues.

He discovered that only those polypeptides made up of

amino acids from natural sources were attacked by the enzymes. This proved that the shape of the artificial amino acids was unsuitable for the enzymes. Again, his lock-and-key analogy held good.

Further light on the way in which enzymes work was shed by a British researcher, A. Brown. In the early 1900s Brown was studying the rate at which an enzyme could act in its substrate. At first, he knew, adding more substrate speeded up the reaction. After a while, however, adding more substrate made no difference. The reaction rate remained steady at what seemed its fastest possible speed.

Brown explained this phenomenon by the hypothesis that the enzyme molecules present in the reaction mixture could do only a certain amount of work at a time. It was as though they were combining temporarily with the substrate.

This idea also explained an observation made by other researchers about the enzyme invertase. They had noted that this enzyme was more chemically stable, in that it could withstand heat better, when it was in the presence of its substrate. This led to the supposition that the enzyme was somehow in a different condition when with its substrate. Brown pointed out that this would be true of the enzyme if it were temporarily combined with its substrate.

Brown's idea about how an enzyme works reinforced Fischer's original lock-and-key concept. Later work has shown that it is also beautifully borne out by a chemical phenomenon called competitive inhibition.

Suppose we have a solution containing succinic acid and an enzyme which acts upon it, succinic dehydrogenase. If we now add malonic acid to the solution, we find that the effectiveness of this enzyme's action on the succinic acid is cut down. Why is this?

Experiments have shown that malonic acid, which has a structure similar to that of succinic acid, attaches itself to the enzyme, so preventing the enzyme from interacting with the succinic acid. The structural similarity of the malonic acid "fools" the enzyme.

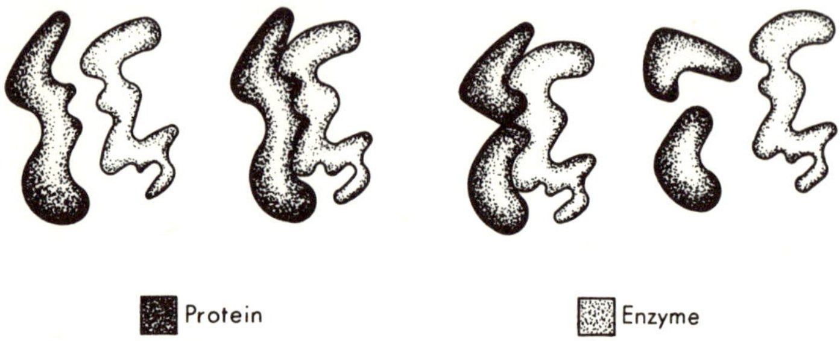

Figure 12. Representation of how a substrate is thought to combine with its enzyme by fitting together. The action of the enzyme then splits the substrate apart. Read the other way, the same scheme portrays the joining together of two molecules under the influence of the enzyme into which they fit.

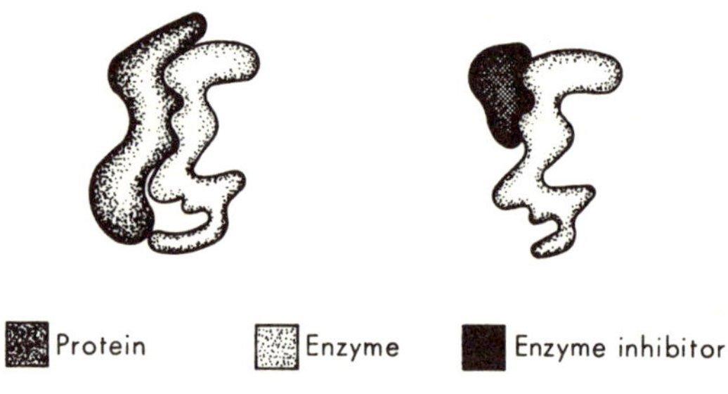

Figure 13. A substance similar in shape to the substrate can block the action of the enzyme. This is competitive inhibition.

Competitive inhibition is turned to good use with sulfanilamide, a drug that inhibits the growth of some disease bacteria. These bacteria normally use as food a member of the B group of vitamins called *para*-aminobenzoic acid. Sulfanilamide has a very similar chemical structure:

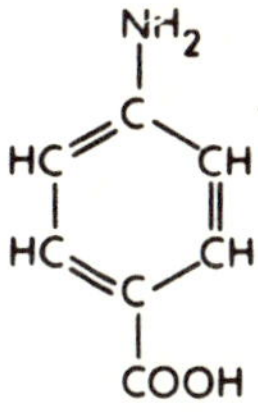

p-aminobenzoic acid

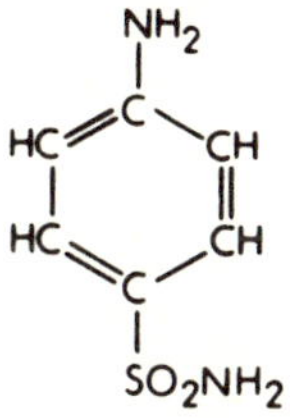

sulfanilamide

The drug thus blocks the bacteria from their effective supply of the vitamin and they starve to death.

Exact analysis of the physical structure of an enzyme molecule had to wait for the emergence of a twentieth-century tool—X-ray diffraction analysis. If a beam of X rays shines through crystallized protein, in which the molecules are regularly arranged, it is split up, or diffracted, in a way that depends upon the arrangement of the atoms in the molecules. This is because the wave length of the X rays is much the same as the distance between adjoining atoms in the crystal.

By letting the diffracted beam fall upon a photographic plate we obtain a complex pattern of light and dark, the light areas indicating where the X rays have struck the developed plate. From this pattern we can work backward and figure out how the beam has been diffracted, and consequently how the atoms are arranged. High-speed electronic computers are needed to perform this analysis.

In this way David Chilton Phillips and his colleagues at the Royal Institution in London, England, worked out the structure of an enzyme for the first time in 1965. The technique that they used had already been applied by Max Ferdinand Perutz at Cambridge in working out the physical structure of hemoglobin, and later by John Cowdrey Kendrew of the same university in analyzing the physical structure of the muscle protein myoglobin.

The enzyme studied at the Royal Institution was lysozyme,

which is found in tears, other body fluids, and egg white. It has a protective function against bacteria because it attacks a sugar found in bacterial cell walls. The molecule contains 129 amino acid residues, the sequence of which had previously been determined by Pierre Jollès and his colleagues at the University of Paris and by Robert E. Canfield of the Columbia University College of Physicians and Surgeons. This information enabled the Royal Institution workers to build a rough model of the molecule only three days after they had completed a map of the way X rays were reflected from different parts of the molecule. A reasonably accurate model was then built over the next six months.

Figure 14. Shape of the lysozyme molecule as shown by a relatively crude model. The substrate of the molecule fits neatly into the deep cleft between two "wings."

Parts of the amino acid chain are coiled around like a wire spring, parts are fully extended, and in one place an extended chain doubles back on itself. The molecule as a whole has two wings with a deep cleft between them.

Phillips and his co-workers were able to identify the active site on the enzyme molecule–the point at which it actually interacts with the bacterial sugar. They found that the cleft fits neatly over the sugar molecule–rather like a mouth biting upon a wafer. Thus they revealed exactly the kind of lock-and-key mechanism prophesied by Fischer.

ENZYMES IN THE CELL

Marvelous as enzymes are, they mostly do not function in isolation but in definite compartments within the living cell. Their activity varies widely and enables cells to carry out different functions for the body. Thus their activity is different in liver cells, muscle cells, and bone marrow cells.

A single liver cell contains around 50,000 million protein molecules. Even if we suppose that only one protein molecule in every thousand is an enzyme, that still makes a total of 50 million enzyme molecules in just one cell.

In the human body there are some 60,000 *billion* cells, all of which must work harmoniously together for the body to be healthy. The orderly regulation of these cells is, substantially, regulation of their enzymes. The mechanism for this control must be transmitted from one generation to another and hence must be packaged in the genes and reside in the cell. To understand the functioning of the genes, then, requires an understanding of their chemical regulation of the enzymes.

The next step in men's knowledge of gene functioning was, in fact, to elucidate just this area of chemical control.

5

"One Gene, One Enzyme"

FOR PROBABLY THREE HUNDRED YEARS, MEDICAL MEN HAVE known of a rare hereditary disease in which the principal symptom is that the affected person's urine turns black on exposure to air. The biochemical basis of this phenomenon was worked out in 1859 by a German scientist named C. Bödeker, who found that the darkening is due to a substance called alcapton, or homogentisic acid. After that, the disease became known as alcaptonuria, meaning "alcapton in the urine."

Alcaptonuria is inherited in a typically Mendelian fashion, just as blue eyes are, and we now know that it results from the presence of a defective gene. That this might be so was first suggested in 1908 by Archibald Garrod, a British physician and biochemist working in Oxford. Garrod also put forward the idea that alcaptonurics might lack an enzyme normally responsible for catalyzing one stage in the body's breakdown of homogentisic acid, so that this accumulated and was excreted. We now know that this hypothesis was correct: an alcaptonuric lacks an enzyme called homogentisic oxidase in his liver.

The special interest of this obscure disease lies in the fact that it is a simple example of what can happen to the body's chemistry when a gene is defective. Clearly, the normal gene is required for the production of homogentisic oxidase.

A more important instance of what Garrod termed an "inborn error of metabolism" is phenylketonuria. In this disease phenylpyruvic acid is excreted in the urine. More seriously, the disease is invariably associated with extreme feeble-mindedness. In the 1930s scientists found that this disease affected members of the same family and its inheritance could be explained as due to the inheritance of a single gene.

The biochemical basis of phenylketonuria lies in the metabolism of phenylalanine, an amino acid that is essential in our diet because we cannot synthesize it. Normally part of this phenylalanine is hydroxylated to tyrosine, another amino acid, but if this change is blocked it is instead oxidized to phenylpyruvic acid. The accumulation of this acid appears to be responsible for damage to the brain, so causing mental deficiency.

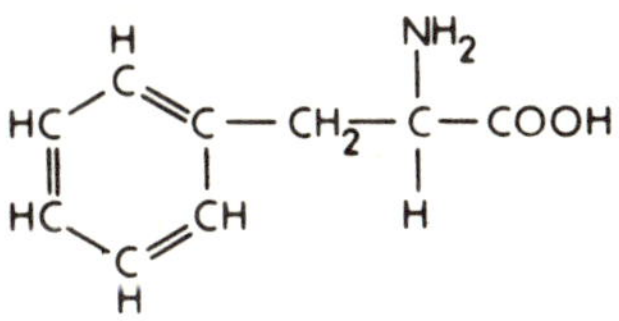

phenylalanine

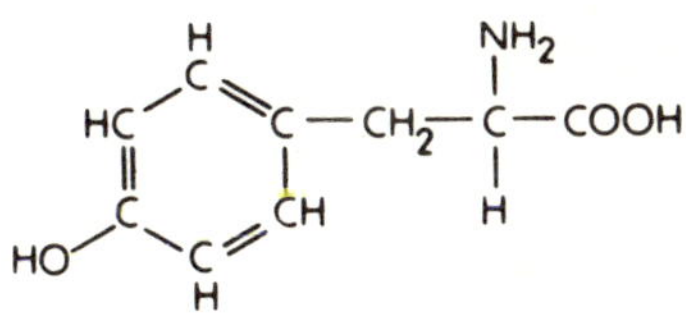

tyrosine

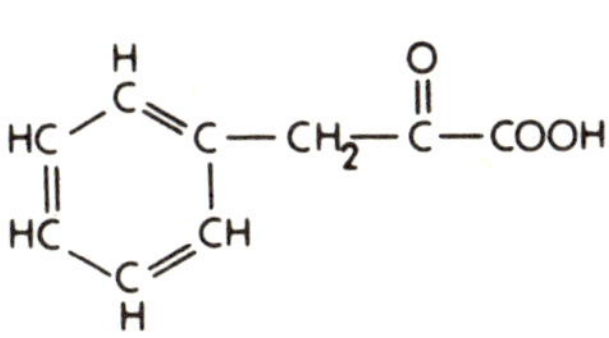

phenylpyruvic acid

In treating this disease, phenylketonuric children are now reared from birth on a carefully prepared diet containing just enough phenylalanine for protein synthesis but too little to cause toxic building up of phenylpyruvic acid.

A third example of an inborn error of metabolism is galactosemia. This rare disease involves an inherited inability to carry out a biochemical reaction needed to use up the galactose sugar supplied in milk. As a result, the baby's body tissues become flooded with unused galactose, which kills it or permanently damages its organs, including its brain. This tragic inability to use galactose stems again from the lack of a single enzyme. To treat the disease, milk is excluded from the infant's diet.

CHEMISTRY OF CHARACTERISTICS

Knowledge of these diseases gives us a clue as to what genes do and how they do it. More exact knowledge has come from the work of George Wells Beadle and Edward Lawrie Tatum, working with a salmon-pink mold called *Neurospora*, which is often found growing on bread.

Beadle was born in 1903 in Nebraska, where his father ran a small farm. (He still dislikes cream because he had to skim it off endless farm pails!) He first learned of Thomas Hunt Morgan's work on fruit flies while studying at the University of Nebraska, and his imagination was fired immediately. As a graduate assistant at Cornell University, he worked in genetics.

After getting his doctorate at Cornell in 1931, Beadle went to the California Institute of Technology, whither Morgan had moved three years earlier to establish a Biology Division. There, Beadle teamed up with one of Morgan's followers, Alfred Henry Sturtevant, and worked on the genetics of corn and of fruit flies. However, he was dissatisfied with the way in which genetics was then organized, considering that it had become too isolated a study, and he found a like-minded colleague in Boris Ephrussi, who had come over to Cal Tech from Paris in 1933.

Two years later, Beadle followed Ephrussi back to Paris and joined forces with him in his section of the Institut de Biologie Physico-Chimique. There, the two researchers plunged into an investigation of how genes could chemically affect the development of an organism.

Sturtevant, in studies of developing fruit flies, had discovered that if one *X* chromosome was missing from a cell in a female embryo, the sector of tissue resulting from later divisions of that cell remained minus one *X* chromosome and was male in character. The bizarre situation was thus created in which a female fly could have a male portion in her body!

A sex-linked character such as vermilion eye color could be present only in this male sector, so that the logical sequence might be a vermilion eye. However, Sturtevant found that actually the eye color appeared normal—as though some substance had diffused in from the surrounding normal tissue and restored normal eye color.

To study this phenomenon more closely, Ephrussi, with Beadle helping him, decided to try the fantastically delicate business of transplanting eyes from one type of fly to another in the larval stages, before the eye color had developed. Other scientists at the Institut were not encouraging and told them that their chances of success were minuscule. However, they persevered, and one day they found that a transplant had "taken"—they had produced a fly with three eyes! To celebrate, the two scientists spent the day at a sidewalk café.

When Ephrussi and Beadle transplanted vermilion eyes into normal, or so-called wild-type, larvae they found that the transplanted eyes became wild-type. The same was true of another eye color, cinnabar. Evidently, some substances were diffusing into the transplanted eyes, causing them to change to the wild-type color. The question now was: were these substances the same or different for the two types of transplants?

Further experiments demonstrated that two different substances must exist, one being formed from the other. The sequence was:

precursor → v^+ substance → cn^+ substance → pigment

where v^+ substance is a diffusible material capable of changing a vermilion eye into a wild-type eye, and cn^+ substance can do the same for a cinnabar eye.

Beadle and Ephrussi were able to interpret their findings on the basis that the gene causing vermilion eyes blocked the change from v^+ substance to cn^+ substance, leaving the eye vermilion, while the gene causing cinnabar eyes blocked the change of cn^+ substance into wild-type pigment.

"What was later called the one-gene, one-enzyme concept was clearly in our minds at this time, although, as I remember, we did not so designate it," Beadle recounted later.

In the fall of 1937 Beadle moved to Stanford University. Here he was joined by Tatum, who tried to identify the chemistry of the diffusible substances. Eventually, their efforts, coupled with those of Adolf Butenandt, a German organic chemist, unraveled the entire sequence.

CHEMISTRY OF MUTATIONS

This success in understanding the chemical effect of a gene had come only after many years of endeavor. To speed things up, Beadle and Tatum determined to try reversing the procedure—to look for gene mutations that would influence known chemical reactions, thus eliminating the laborious business of identifying a host of unknown substances. What organism could they use for this purpose?

Several factors led to the same decision. While still a graduate student at Cornell, Beadle had heard Bernard Ogilvie Dodge of the New York Botanical Garden give a seminar on inheritance in the bread mold *Neurospora*. Dodge was an enthusiastic supporter of *Neurospora* as an organism for genetic studies.

"It's even better than *Drosophila*," he used to insist to Thomas Hunt Morgan, whose laboratory he often visited.

Finally, he persuaded Morgan to take some *Neurospora* cultures with him when he moved from Columbia University to Cal Tech in 1928. Shortly afterward, Carl Clarence Lindegren came to Morgan's laboratory as a graduate student and worked out much of the genetics of *Neurospora* for his thesis. Beadle and Tatum had this valuable knowledge ready to hand.

Further, Tatum had spent a post-doctoral year at the University of Utrecht before joining Beadle, and had worked in the laboratory of F. Kögl, who had just discovered biotin, one of the vitamin B group. Now it so happens that the diet needed by *Neurospora* is simply various inorganic salts, sucrose (sugar), and biotin. From this minimal diet *Neurospora* can synthesize all of the twenty amino acids it needs, and all of its vitamins—except, of course, biotin. (It is a curious fact that so-called simple organisms, such as *Neurospora*, can perform prodigies of chemical synthesis whereas "advanced" organisms, such as ourselves, have lost the ability to make many of these substances and consequently must have them supplied in our food.)

A year earlier, Beadle and Tatum could not have used *Neurospora* for their purposes, because not enough biotin would have been available to feed their cultures. As it was, however, the mold was ideal for their needs. Their strategy was simple: because *Neurospora* can synthesize so many substances, it must contain genes to govern the chemical sequences involved. They figured that, by causing some of the mold's genes to mutate, they should soon produce molds with specific "inborn errors of metabolism" (to use Garrod's phrase)—errors which they could identify chemically without too much difficulty.

Describing their work later, Beadle explained their approach by an analogy. By watching the cars emerge from an automobile plant, it would not be possible to determine what each worker did in assembling them. However, if one could replace able workers with defective ones, and then observe the result in the product, it would be possible to determine that one worker put on the radiator grill, another added the carburetor, and so on.

How were Beadle and Tatum to set about making the mold's "workers"—its genes—defective? Fortunately, a suitable technique had been evolved earlier by Hermann Joseph Muller at the University of Texas. In 1927 Muller proved that irradiation of sex cells with X rays causes hereditable mutations. For this pioneering work, he received the Nobel Prize in Medicine and Physiology in 1946. After Muller's work, other mutagenic agents—as they are called—were discovered, including ultraviolet rays. Beadle and Tatum had a whole armory of mutagenic agents at their disposal.

Like many simple organisms, *Neurospora* reproduces both sexually and asexually in its life cycle. In the asexual stage, it multiplies without any genetic change, so that unlimited quantities of pure-breeding examples of the mold were available for Beadle and Tatum. In the sexual stage, there is a fusion between nuclei from molds of opposite sex reactions. Each fusion nucleus then divides to produce eight spores neatly arranged in line in a spore sac.

If the two parental strains differ by only a single gene, four of these spores will always carry descendants of one form of the gene, while the other four spores will carry descendants of the other form. A skilled laboratory worker, using a microscope, can dissect these eight spores in sequence from their sac and plant each separately in a culture tube. By adjusting the food requirements in the culture tubes, he can determine any changes in the dietary requirements of the mold spores.

The procedure adopted by Beadle and Tatum was therefore as follows. First, they treated asexual spores of the mold with a mutagenic agent and allowed these to reproduce sexually, producing spore sacs. Each of the eight spores in each sac was then planted separately on a nutritional medium containing everything the mold could need for growth, including substances which it would normally manufacture for itself.

Thus far, any deficiency that the mold might have in making any of its food substances because of some induced mutation could not show up, because of the richness of the nutritional medium.

Now samples of the luxuriantly growing mold were taken from each of the eight cultures and placed on the basic medium needed by the normal mold. If a harmful mutation had occurred, Beadle and Tatum reasoned, four of the eight samples would not grow on this minimal medium because of the resulting biochemical defect.

Once they had found that a defect existed, the researchers had to try out further samples on other culture media, to each of which some substance had been added from the range which the mold would normally make. That way, they would eventually find one such culture medium which would allow the mold to grow—the one containing that added substance which the mutant mold could no longer make.

In this painstaking way, Beadle and Tatum began to check

different irradiated molds. Their first mutant came with the 299th, which required vitamin B_6 in addition to the usual minimal diet, and the second came with the 1,085th, which required vitamin B_1. They vowed to keep going until they had ten different mutants, but to their gratification they soon had dozens.

ONE GENE, ONE ENZYME

Having found mutant molds with definite biochemical defects, Beadle and Tatum needed to determine whether each of these defects was the result of mutation in only one gene or in more than one. Beadle wrote later:

> It was a simple matter to determine whether our newly induced deficiencies were the result of mutations in single genes. If they were, crosses with the original should yield four mutant and four nonmutant spores in each spore sac. They did.

By these studies, Beadle and Tatum were able to demonstrate that each gene in the mold controlled a specific chemical synthesis. Since enzymes are required for chemical syntheses, they figured that each gene probably controlled a specific enzyme; this was to become famous as the one-gene, one-enzyme concept. For their work on this, Beadle and Tatum were awarded a share of the 1958 Nobel Prize for Medicine and Physiology.

Many other workers took up the thread of research after Beadle and Tatum reported their first results in 1941 and 1942. They soon showed that quite a number of genes could be involved in the synthesis of any substance.

For example, three genetically different mutants of *Neurospora* were created which could not synthesize arginine, an amino acid. However, not all of these mutants needed to have arginine added to their basic diet in order to flourish. Of the three, one required arginine, one arginine or another amino acid called citrulline, and one arginine or citrulline or a third amino acid called ornithine.

From these data it appears that *Neurospora* can normally start by making ornithine, convert this to citrulline, and then change this to arginine.

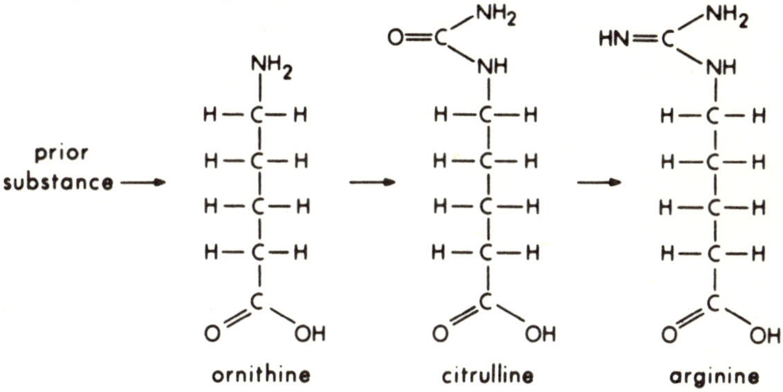

From this, researchers argued that each link in this chemical chain was forged by a specific enzyme and this enzyme was genetically controlled. Of course, we need not suppose that just one enzyme is needed for each of these three stages of synthesis: other intermediate compounds may exist. Further, more than one gene may be involved at each stage, if such genes are all affected by the mutation—as they might be if located sufficiently near each other. None of this, however, invalidates Beadle and Tatum's hypothesis that a single enzyme is controlled by a single gene.

ONE GENE, ONE POLYPEPTIDE CHAIN

A closer view of the control exercised by a single gene has come from work connected with a hereditary disease known as sickle-cell anemia. This abnormality exists among peoples native to northern Africa and the Mediterranean countries of southern Europe. It is so severe a disorder that the affected persons usually die young.

Under the microscope, the red blood cells of an affected person appear as odd crescent or sickle shapes instead of the normal round cells. The functioning of these cells is defective: the hemoglobin they contain does not carry as much oxygen as it should, and so too little oxygen reaches the body tissues.

The genetic pattern of this disease was worked out in 1949 by James Van Gundia Neel at the University of Michigan. He showed that the anemia occurs when both genes in a pair are defective. If only one gene in the pair is defective, the result is a much more common condition known as the sickle-cell trait. This causes little or no interference with normal activities.

Linus Pauling and co-workers at Cal Tech discovered that there is a chemical difference between normal and sickle-cell hemoglobin. To do this, they used a technique called electrophoresis. When normal hemoglobin was placed on a paper soaked in a suitable conducting solution and exposed to an electric field, the hemoglobin moved toward the positive pole. On the other hand, hemoglobin from a person with sickle-cell anemia moved toward the negative pole. Hemoglobin from someone with the sickle-cell trait divided, moving toward each pole, so behaving like a mixture of normal and abnormal hemoglobin. Pauling showed that the differences between the two hemoglobins lay in the protein part of the molecule, the globin, for the heme (iron-bearing) portions were identical.

The exact chemical difference between these two versions of the globin was eventually worked out by Vernon M. Ingram at Cambridge University in England. To do this, he used the digestive enzyme trypsin to break the hemoglobin molecule down into peptide chains, each containing some ten amino acids. He obtained twenty-eight different peptides and proceeded to analyze these by a mixture of two techniques: paper electrophoresis and paper chromatography.

First, a drop of solution containing the peptides was deposited near the edge of a large sheet of moist filter paper immersed in a special solution. An electric current was then passed through the solution in the paper. Under its influence, each peptide migrated through the paper and as each traveled at its own speed the different peptides separated from each other and formed a line of spots.

Now the sheet was dried and hung so that the line of spots touched a special liquid. In the course of several hours this liquid rose along the thin fibers of the paper, carrying the peptides from each spot with it. (This is the technique used in paper chromatography.) Because each peptide migrated upward at its own rate, the peptides were separated into a network of twenty-eight spots on the paper.

"I called this map a 'fingerprint' of the hemoglobin," Ingram recounted later.

Using this technique with normal and then sickle-cell hemoglobins, Ingram obtained "fingerprints" for each. On comparing the two, he found that they were identical except for one peptide.

Now he had to analyze this errant peptide chemically. This was a very tedious business, for to obtain enough peptide (a few thousandths of a gram) Ingram had to make many such "fingerprints" and cut out the piece of paper on each that bore the differing peptide. From this piece of paper he obtained impure peptide, and then he had to purify it.

The analysis revealed that the peptides were identical except for one amino acid. Whereas the normal peptide had two glutamic acid residues and one valine residue, the sickle-cell version had a single glutamic acid residue and two residues of valine. This replacement of glutamic acid by valine was the sole change, yet it was enough to confer a grave disease upon the unfortunate possessor! Out of a chain of some three hundred amino acids in the whole hemoglobin molecule, one was different—and it was different because of a change in a single gene.

Studies such as this altered the one-gene, one-enzyme concept to a one-gene, one-polypeptide-chain concept. Whether or not gene mutation can influence protein structure in other ways, we now know that it can act by affecting the sequence of a single amino acid.

Here, then, was striking evidence for the way in which a gene acts by exercising chemical control over the body's functioning. The next question to be solved was: what is the chemistry of the gene itself?

6

The Chemistry of the Gene

Deoxyribonucleic acid was discovered in the nucleus of pus cells by the Swiss biochemist Friedrich Miescher in 1869. Although long neglected, the fact that this substance was present only in the nucleus of cells, the controlling center and the resting place of the chromosomes, was undoubtedly suggestive. Could DNA, as it came to be called, form an essential component of the chromosomes?

The DNA in the nucleus is in fact present in the chromosomes and not in the rest of the nucleus. That this is so was shown in 1924 by a German biochemist, Robert Feulgen. He had discovered that when DNA is warmed with strong acid and then treated with a dye, acid fuchsin, it turns a brilliant crimson. When he stained sections of tissues in this way, he discovered that only the chromosomes turned crimson, proving that they alone harbored DNA. Soon it was found that plant and animal cells alike have DNA only in the chromosomes.

The Feulgen test was applied to the giant salivary chromosomes of *Drosophila*. It revealed the fact that the bands on the chromosomes are particularly rich in DNA. Did this mean that DNA was the actual material of the gene, the basis of heredity?

In the late 1930s a group of biologists in Sweden, working under Torbjörn Caspersson, began examining cells under ultraviolet light. Nucleic acids absorb this light very strongly, and Caspersson devised a way of measuring this absorption and using the result to deduce how much nucleic acid was present.

By using photoelectric cells and amplifiers of great sensitivity, Caspersson was able to plot the degree to which different cell components absorbed ultraviolet light. He demonstrated that nucleic acids are present in greater amounts than usual wherever cells or tissues are growing rapidly. This suggested that nucleic acids must have an important function in the synthesis of new cellular material.

Around 1950, two groups of investigators independently set out to measure the relative amounts of DNA in different types of cells. They were André Boivin and his collaborators at Strasbourg in France, and Alfred Ezra Mirsky and Hans Ris at the Rockefeller Institute for Medical Research in New York.

In one experiment Mirsky began by counting the number of red blood cells in a certain volume of blood taken from a rooster. He then measured the amount of DNA in that same volume and divided it by the number of red blood cells to arrive at the amount of DNA per cell. The figure was 2.3 hundred-millionths of a milligram.

To analyze the DNA content of cells in solid tissues such as liver, spleen, or kidney, the researchers first dispersed the cells in a fluid so as to be able to count them. For example, liver was immersed in citric acid and then disintegrated in a high-speed mixer. Although the cells were broken down in this treatment, the nuclei remained intact. The researchers then separated these nuclei from the rest of the cell debris by spinning the mixture in a centrifuge. Being heavier than the debris, the nuclei settled at the bottom of the tube. Once the rest of the mixture had been drained off, the clean nuclei were dispersed in citric acid and counted. After this, the quantity of DNA per nucleus could be figured in the same way as for the red blood cells.

The researchers analyzed a variety of tissues in this way. The significant finding they reported was that all types of body cells in any particular species of animal had the same characteristic quantity of DNA. Thus the frog has 15 hundred-millionths of a milligram in every body cell nucleus; the shad has 2 hundred-millionths; and the green turtle has 5.3 hundred-millionths. In the germ cells, however, just half those amounts of DNA are present—and significantly the germ cells carry half as many chromosomes as do the body cells.

All of this indicated that DNA was closely associated with the hereditary factors in the chromosomes. Various proteins also exist in the chromosomes, but they are not distributed in cell nuclei in the same regular proportions.

Still further evidence came from staining experiments performed by Mirsky and Ris. They stained cells with DNA-specific Feulgen stain and then estimated the amount of light absorbed by the stain, using a photoelectric cell and a microscope. The more DNA in a nucleus, the more light was absorbed. Now some nuclei in the body contain two or four sets of chromosome pairs, instead of the usual one; Mirsky and Ris found that these nuclei contained correspondingly greater amounts of DNA.

THE TRANSFORMING FACTOR

In 1928 a British public health officer, Frederick Griffith, made some curious bacteriological observations. He was studying the bacterium which causes pneumonia, *Diplococcus pneumoniae*. There are various types of pneumococcus, and one way of distinguishing between them is by their appearance when grown in colonies on nutritive plates. These colonies are little mounds of the bacteria. Some types of pneumococcus form colonies with a rough surface, and some form colonies with a glistening, smooth surface. The rough, or R, types are harmless, while the smooth, or S, strains are virulent.

Normally, the S strain of pneumococcus is encountered, but Griffith found that by passing this strain on from one mouse to another he could eventually obtain an R form. How could this happen?

The difference between the virulent and the nonvirulent strains is basically that the virulent cells have capsules of a polysaccharide material around them while the others do not. Somehow, the S strain had lost its capsule during its passage through the mice.

Each type of pneumococcus is normally very stable. It can be grown pure for countless generations. However, about one S cell in every ten million will give rise to a colony of R cells. Here again is the mysterious change of the kind noted by Griffith. An abrupt genetic change, or mutation, has occurred.

Griffith found that if he heated the virulent bacteria to 60°C. they proved harmless when injected into mice. He then tried injecting mice with heat-killed virulent bacteria plus live nonvirulent bacteria. To his astonishment, he found that the mice sometimes died of pneumococcal infection. What was more, the blood of these unfortunate animals was always infested with live virulent bacteria—bacteria with capsules!

When isolated and grown in cultures, these bacteria turned out to be identical with the virulent heat-killed bacteria that had been injected—except, of course, for the fact that they were alive. How could this extraordinary transformation have occurred? Evidently, some factor must have passed from the heat-killed virulent bacteria to the living nonvirulent bacteria, causing them to change their nature—but what could this transforming factor be?

Griffith bravely attempted an explanation, but we know today that it was faulty. He suggested that the heat-killed virulent bacteria broke up in the body of the mouse to form a material used as food by the live, nonvirulent bacteria. Then, so he conjectured, the nonvirulent bacteria were transformed by this diet into virulent bacteria.

Other workers confirmed Griffith's results and attempted to explain them. Besides observing the mysterious transformation in the body of a mouse, Martin H. Dawson of the Rockefeller Institute induced the transformation in the test tube. He mixed encapsulated (virulent) heat-killed bacteria with nonencapsulated live bacteria in a nutrient solution in a test tube and left them. Sure enough, he obtained some encapsulated living bacteria.

Another important step was taken in 1932 by James L. Alloway, also at the Rockefeller Institute. If pneumococci are placed in a solution of bile salts in a test tube, they disintegrate. Alloway filtered a solution of heat-killed encapsulated cells which had been disintegrated in this way and found that the filtrate still contained the transforming agent.

From this it was clear that the transforming agent, although normally contained in the bacterial cell, could act independently of it. Hence this agent, which altered the hereditary constitution of bacteria, was nonliving chemical material. It was, in fact, genetic material. The next step—and a momentous one—was to isolate this material and unravel its chemistry.

THE CHEMISTRY OF HEREDITY

The person who undertook to identify the transforming agent was a small, shy, sensitive man named Oswald Theodore

Avery. He was a medical doctor who had chosen research rather than the practice of medicine, and he devoted himself single-mindedly to research throughout his career.

To be as close as possible to his laboratory in the Rockefeller Institute, he lived right across the street. He never married and he rarely took time out to travel. He even stopped answering correspondence because of the precious hours it stole from his research.

Avery had worked on the pneumococcus from as early as 1913. When he heard of Griffith's work, he seized on it at once. At his instigation, Martin Dawson repeated and extended it.

Then, in the early 1940s, Avery enlisted the help of two young research physicians, Colin Munro Macleod and Maclyn McCarty, in his attempt to pin down the transforming agent.

To whittle down the area of uncertainty, the researchers used a technique of selectively attacking the known chemical components of the bacterial cells with enzymes. That way, they hoped by a process of elimination to arrive at the transforming agent.

They therefore attacked the polysaccharide capsule of the S-type cells, using a specific enzyme which had been purified by another of Avery's young associates, René Jules Dubos. After removal of the enzyme and the remains of the polysaccharide, they found that the product was still a potent transforming agent. Evidently the agent was not the polysaccharide.

Similarly, they attacked the protein components of the bacterial cells with protein-splitting enzymes. In this, they were greatly helped by gifts of pure enzymes from two of their colleagues at the Rockefeller Institute, John H. Northrop and Moses Kunitz. These enzymes did not affect the transforming agent either. Hence it was not a protein.

What else could it be? It certainly was not a lipid, because the chemical methods of isolation and purification being used by Avery and his colleagues would have destroyed any lipid. Of the known materials of a cell that remained, the only major kind was the nucleic acids, RNA and DNA.

As they investigated further, the researchers found that the transforming agent did indeed contain phosphorus, just like a nucleic acid. A complete analysis was then made of the final product of Avery's research, and this was contrasted with the known chemical composition of DNA. The two analyses were almost identical.

As a clincher, the Rockefeller workers subjected the solution containing the transforming agent to the action of a purified enzyme known to attack DNA specifically. After this treatment, the transforming agent lost its effectiveness. There was no more room for doubt: the transforming agent was DNA.

In 1944 the three researchers announced the results of their work: "The active transforming material . . . contains no demonstrable protein, unbound lipid, or serologically reactive polysaccharide, and consists principally, if not solely, of a highly polymerized viscous form of deoxyribonucleic acid."

Although Avery did not spell it out, he had in fact isolated the genetic material of the cell. For the first time, a material basis had been found for the mechanism of inheritance so painstakingly studied by Mendel—and it turned out to be that long-neglected substance Miescher had first isolated from pus back in 1869!

DNA IN BACTERIA

Avery's discovery stunned most biological scientists. Despite previous pointers to the importance of DNA, the majority of workers had always assumed protein to be the material of inheritance.

Protein was favored because of the known protein component in chromosomes and, more particularly, because the complexity of protein molecules allows great diversity. Everyone agreed that only a substance which could occur in very many chemical forms was suitable for carrying the immense number of instructions which must be embodied in any genetic material. DNA, on the other hand, was thought of as a simple four-nucleotide molecule, built up in the way originally envisaged by Phoebus A. T. Levene.

Even after Avery's findings, therefore, scientists continued to doubt their meaning. For example, it was suggested that the DNA he had found might simply have stimulated a genetic variation in the pneumococci and that the actual nature of the variation (to the virulent encapsulated form) was a mere coincidence.

Thus Avery's findings had first to be confirmed. Fortunately, he was such a meticulous experimenter that this was no great problem. Conviction of his correctness spread with corresponding speed from laboratory to laboratory.

In 1951 Rollin Douglas Hotchkiss, also working at the Rockefeller Institute, isolated DNA from pneumococci that had developed resistance to the antibiotic penicillin. He found that this DNA, too, could transform the genetic character of other pneumococci, conferring on them a resistance to penicillin which they had not previously possessed. The same was true for DNA from pneumococci that had developed a resistance to another antibiotic, streptomycin.

Hotchkiss added the DNA from antibiotic-resistant bacteria to the broth in which ordinary pneumococci were growing. He found that these pneumococci took from thirty to sixty minutes to incorporate the DNA and show resistance to the antibiotic. After still longer, the pneumococci duplicated the new DNA and so handed on the resistance to subsequent generations. Here was another convincing example of the genetic role of DNA.

By 1955, some thirty different cases of such genetic transformation had been induced in bacteria in this way. However, the genetic power of DNA had been demonstrated only in bacteria. Did it apply to other kinds of living things as well?

DNA IN VIRUSES

This question was strikingly answered by work on a class of virus called bacteriophage, which infect bacteria. Those which have been closely studied include tadpole-shaped T phage viruses, which attack the *Escherichia coli* bacterium, a harmless resident in the human intestine.

When a dilute suspension of this bacteriophage is poured over a continuous culture, or "lawn," of *E. coli* growing on a flat plate of agar jelly, the bacteria dissolve into clear areas where the viruses have infected them. The bacteria have in fact broken open, or lysed.

Until the advent of the electron microscope, with its immensely high magnifications, the mechanism of this phenomenon was not understood. This microscope reveals that the virus has a head in the shape of a hexagonal prism and a long, thin tail from which project six tail fibers. The entire phage is a mere seven millionths of an inch long.

In attacking a bacterium, the viruses become attached to the surface by their tails. About twenty minutes later, a spectacular event occurs: the bacterium bursts and releases a few hundred complete new viruses. The original infecting viruses are still visible in outline attached to the bacterial cell wall but, as Thomas Foxen Anderson of Philadelphia noted, they are empty husks.

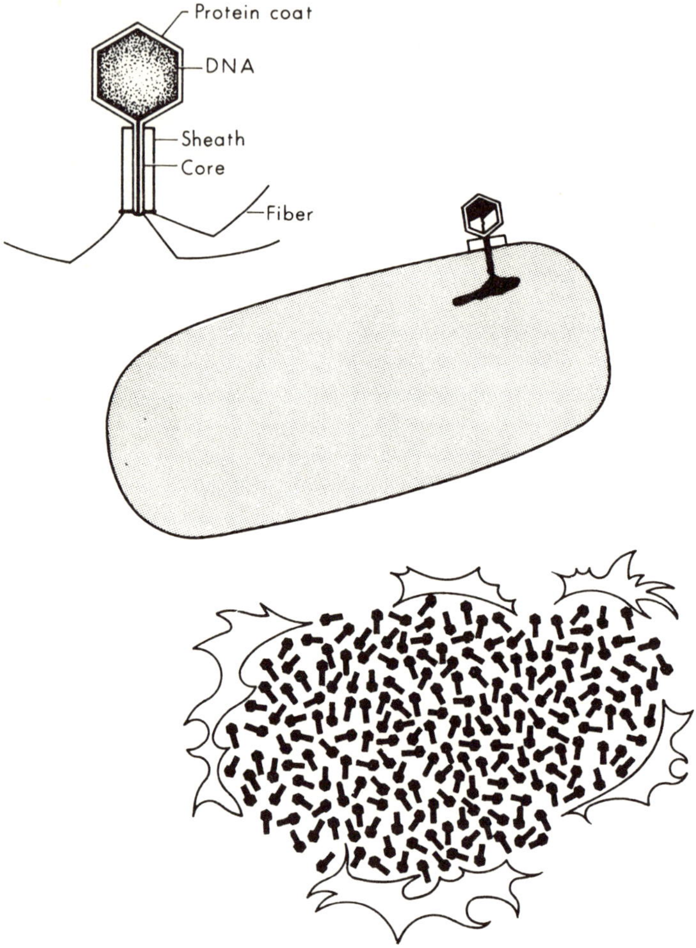

Figure 15. Infection of a bacterium by a T phage. The virus attaches itself by tail fibers to the bacterium wall. Contraction of the sheath then injects the phage content into the bacterium. Some twenty minutes later, the bacterial cell bursts and releases a few hundred new viruses.

The contents of the infecting viruses have evidently been injected into the bacterium, where they have caused the formation of new viruses, made out of the bacterial cell. In fact, the viruses have sent their genes into the bacterium and somehow overcome the normal cellular machinery of the bacterium, causing it to destroy itself in manufacturing more and more viruses.

Once this was known, the question remained: what was the genetic material injected by the viruses?

To answer this, it is necessary to get the contents of the viruses out of their body coats and make a chemical analysis. The contents can be obtained by suspending the viruses in strong salt solution and then rapidly adding water. The resulting change in osmotic pressure forces water into the viruses and bursts them, liberating their contents. By centrifuging off the body coats, the researcher obtains a preparation of the virus contents.

Roger Moss Herriott of Johns Hopkins University analyzed the contents so obtained. He discovered that it was mostly DNA, plus a little protein. Most of the virus protein, however, remained with the coats. This experiment suggested that the virus genes were made of DNA.

To settle the problem, Alfred Day Hershey and Martha Chase performed a truly ingenious experiment in 1952. Working at the Cold Spring Harbor genetics laboratory of the Carnegie Institution, they devised a way of using radioactive isotopes as tracers to follow the movement of the virus protein and DNA respectively. Their strategy depended upon the fact that DNA contains phosphorus but not sulfur, while protein contains sulfur but no phosphorus. The virus DNA can therefore be labeled with a radioactive isotope of phosphorus and the virus protein with a radioactive isotope of sulfur.

To tag the viruses in this way, Hershey and Chase started

by growing *E. coli* bacteria in a medium containing either radioactive sulfur or radioactive phosphorus. The bacteria took up these radioactive elements from the media and incorporated them into themselves. The bacteria were labeled with only one isotope or the other because it is difficult to differentiate between the two when they are present simultaneously.

The labeled bacteria were then infected with viruses and the new viruses formed as a result were found to be labeled in their turn with radioactive phosphorus or radioactive sulfur. Now these tagged viruses were used to infect unlabeled bacteria.

The viruses attached themselves to the bacteria, but after a few minutes Hershey and Chase separated them by stirring the whole suspension in a Waring Blendor (also used in the ordinary kitchen for making milk shakes). They proved that even in those few minutes material had been injected by the viruses into the bacteria, because when a sample of the suspension was allowed to go on incubating, the bacteria burst, liberating new viruses in the usual way.

However, Hershey and Chase did not incubate most of the suspension. Instead, they cooled it in order to stop further incubation of the viruses, and centrifuged it, so separating the bacteria from the viruses. Now they could assay the amounts of radioactive isotope inside the infected bacteria.

When sulfur-labeled viruses were used, the two researchers found that the bacteria were not radioactively labeled. The sulfur-tagged protein of the virus had remained in the coats of the infecting viruses. However, when infection was by phosphorus-tagged viruses, the bacteria were radioactively labeled. This proved that phosphorus-containing DNA had passed into the bacteria, and confirmed its genetic importance.

Today scientists believe that all genes are made of DNA, except for those in some viruses which are made of RNA. They believe this even though experiments of the kind performed by Hershey and Chase obviously cannot be carried out on higher organisms.

One good reason for this belief is the known distribution of DNA in chromosomes, and the halving of its quantity in the germ cells. Another is the remarkable similarity of chemical processes in living cells of different organisms. Thus, in both men and bacteria, individual cells obtain their energy in much the same way. The laws of heredity, too, operate with equal validity for both. Given these facts, why should we suppose that the chemical basis of these laws is not also the same?

7

The Structure of DNA

ONCE DNA HAD BEEN REVEALED BEYOND QUESTION AS THE CHEMical basis of the gene, the immediate problem was to understand its structure. Until Avery's epoch-making work, chemists had been content to envisage nucleic acid as a simple long-chain molecule composed of a repeating sequence of four nucleotides, one of each possible variety. This was termed the "tetranucleotide hypothesis." Such a simple molecule could not possibly embody the complex chemical information required to carry hereditary characteristics. If it consisted merely of a single sequence of four nucleotides repeated over and over, it would hardly be more complex than a molecule of cellulose or starch. Evidently, then, the tetranucleotide hypothesis was inadequate. How should it be corrected?

Work to resolve this problem was undertaken by Erwin Chargaff of Columbia University. He made many very careful analyses of the relative amounts of the four bases occurring in DNA taken from a wide variety of different organisms, such as pig, salmon, turtle, yeast, and viruses. By 1947, he had found that the two purine nucleotides and the two pyrimidine nucleotides were present in widely varying amounts. Hence, there could not be any simple sequence of four nucleotides repeated over and over. He also found that the proportions of purine and pyrimidine differed a great deal from one organism to another: each organism had its own specific proportions. Chargaff therefore knew that the DNA molecule was not only long enough, but varied enough, to be a suitable carrier of genetic information.

So far, so good; but Chargaff also discovered something else, without at that time grasping its significance. He wrote:

> It is, however, noteworthy–whether this is more than accidental, cannot yet be said–that in all deoxypentose nucleic acids examined thus far the molar ratios of total purines to total pyrimidines, and also of adenine to thymine and of guanine to cytosine, were not far from one.

Since thymine and cytosine are pyrimidines, and adenine and guanine are purines, this meant that all DNA molecules contained as many of the large purine bases as of the small pyrimidine bases. Later it was to become apparent that this was a critical feature of the architecture of the DNA molecule.

Meanwhile, the chemical information available told the investigators little about the physical structure of any DNA molecule–that is, how the nucleotides were arranged in three dimensions. Did each molecule consist of one nucleotide chain, of two, or even of three? If two or more chains existed, how were they joined together? Was a nucleotide chain straight or coiled?

The next step in answering these questions was to be through using various physical techniques of investigation–particularly X-ray diffraction.

THE PHYSICAL INVESTIGATIONS

The great pioneer in using X-ray diffraction techniques to examine the structure of long biological molecules was a British investigator, William T. Astbury. While working at the University of Leeds in the early 1930s, he began using X rays to get diffraction pictures of fibrous proteins, particularly keratin, which is the main constituent of wool.

When a wool fiber is heated in water and pulled, it can be stretched, and its X-ray pattern changes too, from what is called the α-pattern to the β-pattern. Evidently, reasoned Astbury, the wool protein was more folded in the α-state, and heating allowed it to be somewhat unfolded and pulled into longer chains. He also noted that many fibrous proteins show the α-pattern, which suggested that these proteins were normally in some partially folded state.

Using X-ray diffraction pictures plus a painfully acquired knowledge of the precise structures of a few amino acids and peptides, Linus Pauling and Robert Brainard Corey were later able to work out the three-dimensional structure of fibrous proteins. They began this work at Cal Tech in 1937 and by 1950 they had the structures of three amino acids and three peptides. Armed with these data, and the rather vague X-ray diffraction pictures obtainable, Pauling and Corey concluded that in the α-pattern the protein was twisted into a helix. (A helix is often loosely called a spiral: thus a spiral spring is really a helical spring.)

A helix is a simple geometrical consequence of joining asymmetric objects together so that each has the same spatial relation to its neighbors. The result is a series of turns progressing along what is called the axis of the helix. Put a series of books in a pile asymmetrically one on top of another, turning each by the same amount relative to the one below, and a helix will result. This rule enables us to understand many helical structures which occur in living systems. In the α-helix, about three and a half amino acid residues are needed to carry the helix forward through one complete turn. More exactly, there are eighteen residues in five turns.

Figure 16. If blocks are placed asymmetrically on top of each other while being turned through a constant amount, they form a helix.

This history of employing X-ray diffraction on proteins, and knowledge of helical structures in nature, were to be valuable in attacking the problem of determining the structure of DNA.

Astbury had been interested in DNA too. In 1938, working with F. O. Bell at University College, London, he took the first X-ray diffraction pictures ever of the substance. He concluded that DNA had a repeating structure in which flat

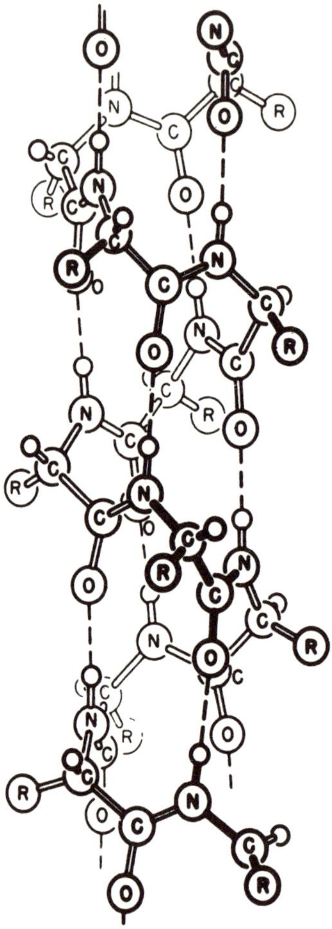

Figure 17. The α-helix of fibrous protein has 3.7 amino acids per turn. In this ball-and-stick model C = carbon, O = oxygen, N = nitrogen, and R = the side chain of the particular amino acid. Hydrogen atoms are represented by small unlettered circles. The dotted lines represent hydrogen bonds.

planes of purine and pyrimidine bases occurred over and over at right angles to the long axis of the molecule. He likened the resulting structure to "a pile of pennies."

However, the main endeavor did not come until 1950 when Maurice Hugh Frederick Wilkins, Rosalind Franklin, and their colleagues at King's College, London, began much more detailed crystallographic studies of DNA. Wilkins, the son of a New Zealand doctor, had been trained as a physicist, and during the Second World War he had worked on the development of the atom bomb. Since then he had turned his energies to the study of life substances, becoming one of the first of a new breed of scientist–the biophysicist, a specialist in the physical exploration of living matter.

When Wilkins, Rosalind Franklin, and their group began work, Mirsky had already found that fibers of a salt of DNA could be prepared very simply by twirling a stir rod in a pure, jellylike solution of the nucleic acid's salt. This was fortunate, because fibers were needed for X-ray crystallography of the material.

The source of DNA Wilkins chose was calf thymus, from which the protein was removed by the standard method of precipitating it out with detergent. Metallic salts of DNA were prepared for the British workers by Leonard Derwent Hamilton at the Sloan-Kettering Institute in New York and were airmailed across the Atlantic as dry material in little cellophane packets. This route circumvented any customs holdup of the precious supplies.

When the DNA was drawn out into fibers, its molecules were oriented in the same direction and consequently they yielded excellent X-ray diffraction photographs. In the crowded basement laboratory at King's College, the investigators made molecular models and fed the measurements from these to an IBM computer in another part of London. There, the data were subjected to the necessary mathematical analysis and compared with the results from the photographs.

The immensely difficult job of Wilkins and his co-workers was to try out all the theoretical arrangements which the models indicated until they found an arrangement which exactly fitted the data from the photographs. In this work, Wilkins showed a remarkable capacity for sustaining very careful observation and thought.

The King's College group eventually found that the purine or pyrimidine bases, each a flat plane in shape, were located a regular 3.4 angstrom units apart. They also found that another regularity in the structure of the molecule existed, being repeated every 34 angstroms along the molecule—that is, at ten-base intervals. They decided that the most likely explanation for this was that the DNA was not straight but helical, forming one complete turn every 34 angstroms.

Further, they showed that the measured density of the DNA molecule could not be accounted for if it was composed of only one nucleotide chain. The density measurements were necessarily crude, because the material they were working with was not all the same—some of the DNA was crystalline, so that its molecules were evenly spaced apart, but some was not. This meant that they could not be sure how many molecules were present in any given weight of the substance. However, their approximate figures indicated that each DNA molecule was composed of between two and three nucleotide chains.

As molecules are normally less closely packed in non-crystalline material, so that the density of the material is correspondingly lower, Rosalind Franklin supposed that fully crystalline material would be denser than the material she was working with. This in turn suggested that the DNA molecule was on the heavy side and composed of three rather than two chains.

Meanwhile, Linus Pauling and Robert B. Corey at Cal

Tech, enlarging on their experience of the helical structure of proteins, put forward the idea that DNA, too, had a helical structure. They believed that this might consist of three nucleotide chains twined around each other with the phosphoric acid groups at the core and the purine and pyrimidine bases projecting outward.

Another clue to DNA structure was provided earlier by a British biochemist named J. M. Gulland, who concluded that the bases must be joined to each other by a very common, weak form of bonding called hydrogen bonding. This idea conflicted with that put forward by Pauling and Corey, for it could only work if the bases pointed inward toward each other.

The position thus far was therefore obscure. The helical structure of DNA was accepted as a working hypothesis but essential questions remained unanswered. Were there two nucleotide chains or three? Did the bases point inward or outward?

THE CROWNING SYNTHESIS

A bold, imaginative synthesis of all the chemical and physical data on DNA was now made by a British researcher, Francis Harry Compton Crick, in collaboration with a young American, James Dewey Watson.

Crick had spent the war years designing mines for the British Admiralty and had then turned to molecular biology. Having worked on X-ray diffraction pictures of the enzyme ribonuclease at the Polytechnic Institute of Brooklyn, he was now back in England, at Cambridge University. There he was joined by Watson, who had a research fellowship. A brilliant, unpredictable man, Watson had been a child prodigy and was a college student at the early age of fifteen. Now he was a biologist.

Early in 1953, the two men decided to collaborate in an attempt to work out the structure of DNA. Recalling this time with Crick, Watson later commented, "He taught me physics. I taught him biology. It worked out rather well."

Their plan of attack was to try building molecular models of DNA until they had one which fitted the known facts. This involved their making exact scale models of the various chemical groups involved, including the lengths and angles of their chemical bonds.

The first point was to decide whether the DNA molecule consisted of three chains or only two. To help themselves, they drew on a further piece of evidence from Wilkins' observations: he had found that the width of the DNA molecule was 20 angstroms.

Armed with this, and assuming that ten nucleotides fitted into every 34 angstroms of length—as indicated by the X-ray diffraction patterns—they were able to figure the density of the molecule from their knowledge of the weights of the individual atoms involved. The result they obtained by assuming one chain of nucleotides was roughly half the known density. Hence they decided that the molecule must have twice as many nucleotides in its length and consist of two chains.

Next, they built models of two nucleotide chains to see how these might fit together. In doing this, they used Chargaff's discovery that in any DNA molecule there were equal numbers of adenine and thymine bases, and of guanine and cytosine bases. To Watson and Crick, this suggested that adenine was somehow paired with thymine and guanine with cytosine. "This pairing is likely to be so fundamental for biology," wrote Crick later, "that I cannot help wondering whether some day an enthusiastic scientist will christen his newborn twins Adenine and Thymine!"

The molecular models the two researchers made elicited the provocative observation that the overall dimensions of each of these pairs were identical. Further, the dimensions of each pair fitted pretty well to the width of 20 angstroms already noted.

With a brilliant flash of insight, the two men used this fact to show how the two bases in each pair could point inward. They decided that the bases were joined together in pairs, forming, as it were, rungs of equal length on a helical ladder. At one stroke, this model accounted for Chargaff's data and explained the regularity of the DNA helix. The mechanism they proposed for the joining of the bases was hydrogen bonding—the same bonding for which Gulland had already provided direct experimental evidence.

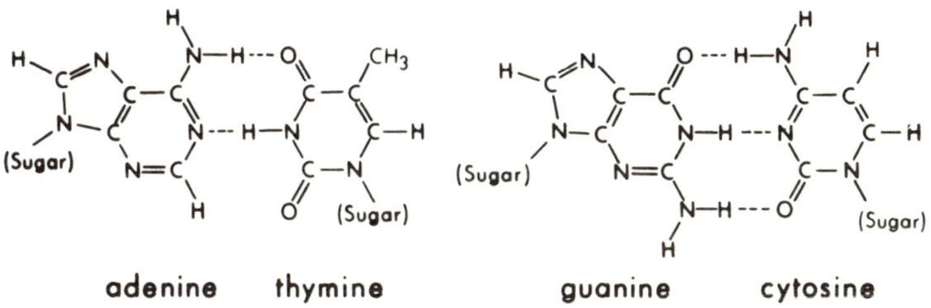

Figure 18. Hydrogen bonding between bases in the DNA molecule. Two bonds link adenine and thymine, three guanine and cytosine. The overall widths of the two pairs are identical.

This work of discussion and model building had occupied only six weeks, during which time Watson had continued his habits of having a leisurely café breakfast each morning and a game of tennis each afternoon. Watson and Crick had succeeded in building a model which revealed that the two chains of nucleotides could twine around each other like the strands of two-ply knitting wool. One chain ran up the helix, and one down: only in this way was a correct fit possible between the bases.

In 1953 Watson, Crick, and Wilkins published their ideas

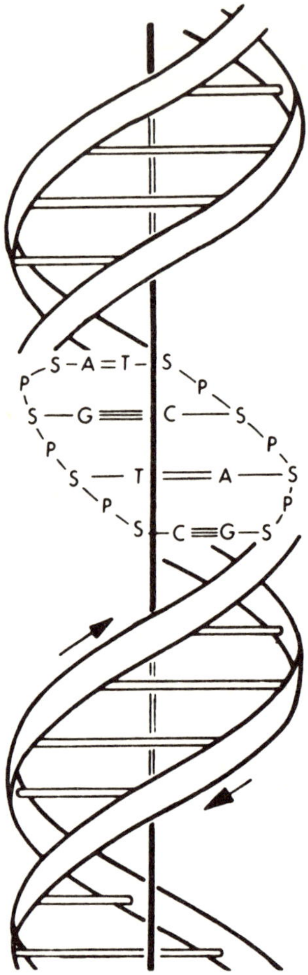

Figure 19. In the Watson-Crick model of DNA the two chains of nucleotides twined around each other like the strands of two-ply knitting wool. One chain ran up the helix and one down.

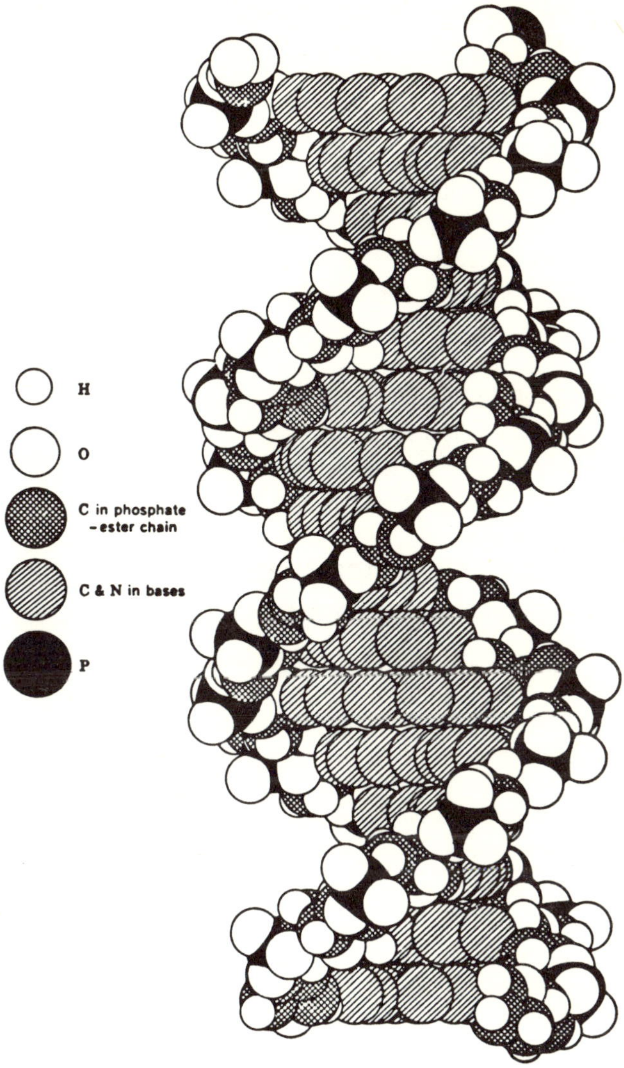

Figure 20. Exact model of DNA by Wilkins to refined data from X-ray crystallography. In this model each atom's sphere of influence is shown solid.

in now-famous articles in the scientific journal *Nature*. The elegance and satisfying simplicity of the Watson-Crick model were at once appreciated by researchers everywhere.

However, the model still corresponded only roughly with the X-ray data. Wilkins and his colleagues therefore began to make further, still more refined observations and obtained pictures of unparalleled excellence. As before, they tested these against measurements obtained from molecular models. After about two more years of unremitting effort they were able to show that the diameter of the double helix envisaged by Watson and Crick had to be decreased somewhat, and that other adjustments were necessary.

For their work on the structure of DNA, Wilkins, Watson, and Crick were awarded the Nobel Prize for Medicine and Physiology in 1962. Crick, who had already built a triumphant copper helix outside his house in the center of Cambridge, now reported, "People keep asking me when I'm going to gild it."

8

The Copying Mechanism

One of the most exciting attributes of the Watson-Crick model was that it at once suggested a mechanism for self-duplication, or replication, of the kind that must occur every time a cell divides in two. "It has not escaped our notice," wrote Watson and Crick in one of their articles in *Nature*, "that the specific pairing we have postulated immediately suggests a possible copying mechanism for the genetic material."

The nucleotide sequence of one chain in the double helix of DNA must have a fixed relation to the other chain, because of the pairing rule. Thus if the sequence in one part of the first chain is ATAGGC (using initial letters to designate adenine, thymine, guanine, and cytosine), the sequence along the second, complementary chain will inevitably be TATCCG. Consequently, if the two nucleotide strands separate, each will serve as a template upon which the other can be built.

Pursuing this idea, Crick suggested that the twin chains of a DNA molecule might begin to unwind and separate in the presence of a supply of new nucleotides. The new nucleotides, he added, would be in the form of preliminary chemical compounds, or precursors, rather than unadorned nucleotides. Then as the two nucleotide chains unwound, loose precursors would attach themselves by their bases to the corresponding bases on each of the single nucleotide chains.

So long as the fit was correct, he argued, each unit would remain attached. Another loose precursor would then fit into place alongside. Gradually a fresh nucleotide chain would grow until a complete complement to the first was formed. Finally, two DNA molecules would exist where there had been only one before.

One snag in this idea was the supposition that the total DNA molecule would unwind into its halves before duplication could begin. As Crick pointed out, the total length of the DNA in a single chromosome must be equivalent to some ten million turns of the helix, although this need not all be in one piece. Still, the imagination boggles at the idea that all this should be unwound before duplication can begin. To get around this, Crick put forward the notion that the synthesis of two new chains might begin as soon as the original ones had started to unwind.

SYNTHETIC DOUBLE HELIX

One of the first pieces of evidence supporting this model came, coincidentally, from quite unrelated research which was being carried out in Severo Ochoa's laboratory in New York University by Marianne Grunberg-Manago. She had come to the laboratory from Paris on a travel fellowship in 1955 and was studying how a compound consisting of adenine, ribose sugar, and two phosphate groups—adenosine diphosphate—is naturally converted into the triphosphate by the addition of another phosphate group. This work was particularly important be-

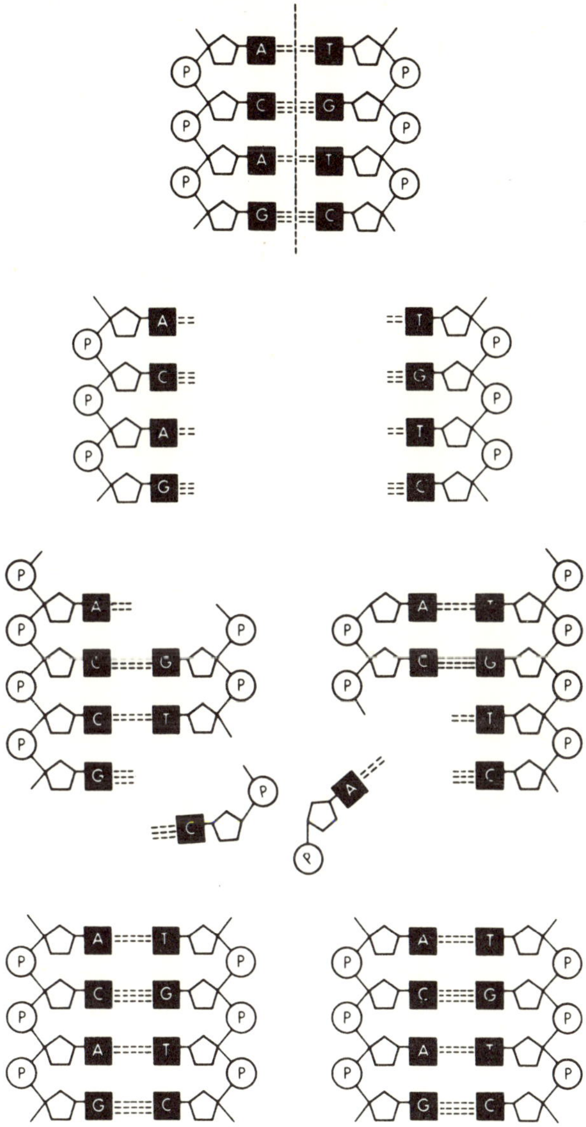

Figure 21. How replication of DNA is believed to occur. The complementary chains separate. Each chain attracts loose precursor units which link to the chains and then to each other. Eventually two complete helixes are formed.

cause adenosine triphosphate, or ATP, is the compound used in the cell to store energy for most of the cell's functions—energy which is released when the third phosphate group is split off.

In her research Marianne Grunberg-Manago used enzyme extracts prepared from a nitrogen-fixing bacterium of the soil, *Azotobacter vinelandii*. She studied the effects of the enzyme extracts on adenosine diphosphate (ADP) by means of paper chromatography. To her surprise, she found that some of the resulting material would not travel through the paper as did a relatively small molecule such as ADP. Instead, it stayed at the beginning edge of the paper. This indicated that it must be a very much larger molecule. Evidently, she reasoned, the ADP had somehow gotten linked together many times over to form a long-chain molecule.

Further work on the material confirmed her supposition. Each of the big molecules was made from many units of ADP by the elimination of one phosphate group. In other words, each unit in the resulting polymer (a large molecule formed from many repeating units) was made of adenine, ribose sugar, and a single phosphate group. But this compound, adenylic acid, is one of the constituents of RNA!

What Marianne Grunberg-Manago had done was to make an RNA-like material in which only one of the four possible bases, adenine, appeared. This material, a polymer of adenylic acid, was christened polyadenylic acid, or poly A.

Ochoa eagerly followed up this work in the hope that this method of synthesis would apply equally well to the formation of real RNA. He discovered that other polymers could be made from each of the other three single bases, repeated over and over. Thus he synthesized poly G, poly C, and poly U with the aid of the enzyme from *Azotobacter*. He was even able to synthesize a truly RNA-like material from all four bases together, thus obtaining a polymer in which they were strung along the polymer at random. For this work Ochoa was awarded a share of the 1959 Nobel Prize in Medicine and Physiology.

We now know that the enzyme which so efficiently couples the diphosphates together, called polynucleotide phosphorylase, is not the enzyme which normally governs the synthesis of RNA in the living cell. However, the fact that it will uncritically join the nucleotide diphosphates together allowed Alexander Rich, at the National Institutes of Health in Bethesda, Maryland, to verify Watson and Crick's belief that the four bases join in certain pairs only.

Rich took some pure poly A in solution and mixed it with a solution of pure poly U. Even as he watched, the mixture became thick and jellylike. What had happened? Such a change indicated that a larger polymer had been formed by some linking between the units of poly A and poly U. X-ray diffraction analysis on fibers drawn from the solution revealed that this was so, and that the strands of poly A and poly U had wrapped around each other. The double-stranded helix so formed had a diffraction pattern very similar to that of DNA.

Thus Watson and Crick's belief that adenine unites with thymine (uracil's equivalent in DNA) was reinforced. Rich's work corroborated their picture of the structure of DNA and of the way in which it replicates.

SYNTHETIC DNA

Direct confirmation that DNA is indeed a self-replicating molecule came from work in 1956 carried out by Arthur Kornberg at Washington University in St. Louis. Kornberg, who had studied under Ochoa, had no less an aim than to synthesize DNA in the test tube, with the aid of the appropriate enzyme.

First he had to obtain a supply of the enzyme. As a source,

he chose *Escherichia coli* bacteria. These divide every twenty minutes and hence, he reasoned, would be rapidly forming DNA. He therefore disintegrated suspensions of these bacteria by bombarding them with high-frequency sound waves, hoping in this way to release the enzyme or enzymes concerned with the synthesis of DNA.

Next, Kornberg made a mixture of the four nucleotides of DNA, plus adenosine triphosphate. He added this because, from knowledge of the energy-releasing role of ATP in living cells, he expected it to activate the nucleotides, which could then combine under the influence of the enzyme.

Unfortunately, Kornberg was not working with a pure extract of the enzyme. All he had was a brew of disintegrated bacteria containing many other enzymes, including ones which would break down any DNA as rapidly as it was formed. Because of this, he could not expect any increase in the total amount of DNA present, even if the enzyme did synthesize molecules of DNA out of his nucleotide mixture. How, then, was he to find out whether any DNA had been made?

To solve this problem, Kornberg hit on the clever idea of labeling one of the nucleotides, deoxythymidylic acid, by incorporating radioactive carbon in it. This radioactive deoxythymidylic acid was supplied by another researcher at St. Louis, Morris Friedkin. Kornberg then incubated the reacting mixture. After allowing enough time for the formation of some DNA, he separated any DNA present by adding acid to precipitate it from the mixture. If this DNA turned out to have any radioactivity, Kornberg would know that some of his labeled nucleotide had been incorporated in it.

He washed the precipitated DNA thoroughly to remove any of the radioactivity of the starting mixture. Then he checked the radioactivity of the remaining pure DNA with a Geiger counter. Just as he had hoped, the counter recorded radioactivity! True, it was only a minute amount—about one twenty-thousandth of the radioactivity left over in the remainder of the mixture—but it was a beginning. He had synthesized a tiny amount of DNA.

Kornberg realized that he needed to get a purer enzyme extract if he was to improve his results. So far, he had opened the door a crack. "Through this crack we tried to drive a wedge," he said later, "and the hammer was enzyme purification."

He achieved this purification by subdividing the various proteins from the disintegrated bacteria into a number of fractions. Eventually he came upon a highly effective DNA-synthesizing fraction, which was several thousand times better than his first crude extract. Kornberg named his purified enzyme DNA polymerase.

Using this purified enzyme, Kornberg experimented with mixtures of the four nucleotides. Instead of adding ATP to supply energy, in this later work he incorporated two extra phosphate groups in each nucleotide, where they were attached by energy-rich bonds. By splitting these off, Kornberg figured, the DNA polymerase would obtain the necessary energy for synthesis.

However, Kornberg discovered that the enzyme would catalyze the synthesis of DNA only if a little DNA was first added to the nucleotide mixture as a primer. Further, he found that any new DNA formed was always an exact replica of this primer DNA, no matter what the relative proportions of nucleotides in his mixture.

Kornberg was now able to synthesize DNA for quite different organisms by use of the appropriate primer, even though the enzyme involved came from *Escherichia coli.* If he used primer DNA from cattle, he made cattle DNA; if he used primer from another bacterium, he made the DNA from that bacterium. Thus he proved conclusively that, with the aid of an enzyme, DNA is a self-copying molecule.

For his work on the synthesis of DNA, Kornberg was awarded a share of the 1959 Nobel Prize in Medicine and Physiology.

REPLICATION OF DNA IN CHROMOSOMES

The work by Kornberg confirmed that DNA is a self-replicating molecule. But were Watson and Crick correct in envisaging that the molecule split in two down the middle? Or could the original DNA molecule produce a copy of itself without doing this first?

One experiment which helped to answer these questions was performed in 1957 by James Herbert Taylor and his associates at Columbia University. These researchers studied the way in which freshly made DNA is distributed in the chromosomes of dividing cells, to see whether it followed the pattern that would be prophesied from the Watson-Crick model.

The DNA in each chromosome replicates before cell division. When the chromosome splits into two daughter chromosomes during the process of cell division, the DNA present is equally shared between the daughter chromosomes.

If the Watson-Crick hypothesis is correct, the original DNA will divide in two during replication and each daughter chromosome will therefore contain half of it. If, on the other hand, the DNA molecule replicates without dividing, one daughter chromosome will contain only old DNA, while the other will contain only new.

To discover which possibility is correct, Taylor and his co-workers determined to use a radioactive tracer technique to follow the progress of the DNA. Instead of using radioactive carbon, as Kornberg had done, they chose radioactive hydrogen (tritium) which was incorporated into thymidine.

They immersed the root tips of seedlings of the bean

Vicia faba in a medium containing this radioactive thymidine. The thymidine was taken up by the seedlings and incorporated into the DNA of their chromosomes as deoxythymidylic acid.

Taylor left the seedlings in the radioactive medium long enough for the chromosomes of many cells to divide once but not again. After removing the seedlings, he washed them free of any radioactivity and then waited a few hours to allow some of the daughter chromosomes to become visible. The question now was: did both daughter chromosomes of a pair contain radioactive DNA, as would be expected from the Watson-Crick model, or did only one member of each pair?

To find out, Taylor used the technique of autoradiography, previously invented in Britain. He made thin sections

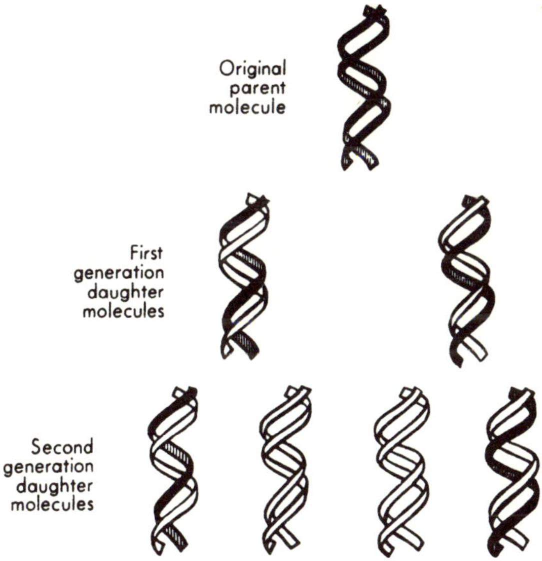

Figure 22. How old and new DNA chains will be distributed if the molecule divides longitudinally during each cell division.

of the root-tip cells on microscope slides and, in the dark, placed photographic film against them. The radioactive tritium in the labeled DNA was emitting beta rays which affected the film. When Taylor developed it, he saw black dots where the rays had struck it. He now took an ordinary photograph of the same cells, so as to show their chromosomes, and superimposed on this the film with dots from the beta rays.

If a chromosome contained labeled DNA, Taylor could now see its outline scattered with black dots. He found that where daughter chromosomes contained labeled DNA, both members of a pair had equal amounts. So far, it seemed that the Watson-Crick model was correct.

As a further check, Taylor extended his experiment. If a daughter chromosome containing DNA which is half radioactive and half not divides, one of its daughters will receive radioactive DNA and the other will not, according to the Watson-Crick model.

Taylor therefore repeated his first experiment and then removed the seedlings and placed them in another nutritive medium in which there was no radioactive thymidine. He left the seedlings there long enough for the radioactive daughter chromosomes to divide yet again and then examined them as before. He found that almost all pairs which were formed from a radioactive parent were made up of one radioactive chromosome and one normal one. This result was in excellent agreement with the predictions made by Watson and Crick.

DIVISION OF DNA IN BACTERIA

Yet another check on the Watson-Crick hypothesis was made in a highly elegant experiment devised by Matthew Stanley Meselson and Franklin William Stahl at Cal Tech and reported in 1958.

The two researchers began by growing *Escherichia coli* in a culture solution containing glucose, mineral salts without nitrogen, and ammonium chloride (NH_4Cl) in which all the nitrogen was heavy nitrogen, N-15, instead of the usual N-14. After allowing the bacteria to divide in two in this medium fourteen times, Meselson and Stahl had bacteria with heavy nitrogen in their DNA. They then abruptly switched the medium by adding a great excess of ordinary ammonium chloride, so that in future growth the DNA formed by the bacteria would have ordinary nitrogen in it.

If the Watson-Crick model were correct, they reasoned, all the bacteria formed in the first division after the addition of ordinary ammonium chloride would have molecules made of half of the old heavy nitrogen and half of the new ordinary nitrogen. At the second division, half of the offspring would contain some heavy nitrogen while some would have only ordinary nitrogen.

To find out whether the DNA was distributed in this way or not. Meselson and Stahl had to check the weights of DNA at successive divisions of the bacteria to discover if the DNA molecules contained heavy nitrogen. Their weighing procedure had to be so delicate that it would register the component of heavy atoms and detect N-15 as opposed to N-14.

To achieve this fantastic degree of accuracy, Meselson and Stahl used a very ingenious technique invented by Jerome Vinograd of Cal Tech. A solution of a heavy salt such as cesium chloride can be rotated in an ultracentrifuge at extremely high speeds which are equivalent to over one hundred thousand times the force of gravity. When this happens, the cesium chloride molecules are thrown toward the bottom (outermost) part of the tube containing the solution. However, the molecules also have a tendency to make their way up the tube (toward the center of the centrifuge) because of random movement due to their heat energy. Thus a balance develops between two forces—the centrifugal and the heat energy. As a result, a gradient in concentration of molecules is set up, there being most molecules at the bottom of the tube and fewest at the top. This also means that there will be a density gradient, the solution being most dense at the bottom and least at the top.

If large DNA molecules are present in this solution, they will move to that level in the tube where the density of the cesium chloride solution exactly matches their own. Eventually the DNA molecules will form a narrow band at this point. The position of this band can be photographed, as the tube rotates, by means of ultraviolet light. This light is strongly absorbed by the DNA, which therefore shows up as a black band in the photograph.

Meselson and Stahl used this technique to discover the density of DNA extracted from their bacteria at different times, before and after the addition of ordinary ammonium chloride to the culture medium.

Their first sample came from bacteria just before the addition of ordinary ammonium chloride to the culture medium. It therefore comprised virtually nothing but heavy DNA, which showed up as one band. The next sample came from bacteria which had divided once after the addition of ordinary ammonium chloride. This also showed one band but higher up the tube at a lesser density. The third sample came after two bacterial divisions and comprised both the DNA of the second sample and an even lighter DNA which was of ordinary density. These two weights of DNA were present in equal amounts. Finally, after four divisions there was a trace of the intermediate weight of DNA but most of the DNA was of the ordinary density.

Meselson and Stahl regarded these results as an exact confirmation of the Watson-Crick model. They interpreted the intermediate-weight DNA as formed of one strand containing heavy nitrogen and one containing ordinary nitrogen.

To double-check, they boiled a solution of the intermediate-density DNA. This relatively mild treatment was enough to break the hydrogen bonds between the bases, thus splitting each molecule into its two constituent chains. When centrifuged, this split-up material no longer showed a band at intermediate density but two sharply distinct ones, the lower at heavy-DNA position, the upper at the position for ordinary DNA. Here was evidence which revealed beyond doubt that DNA replicates by dividing down the middle into two halves.

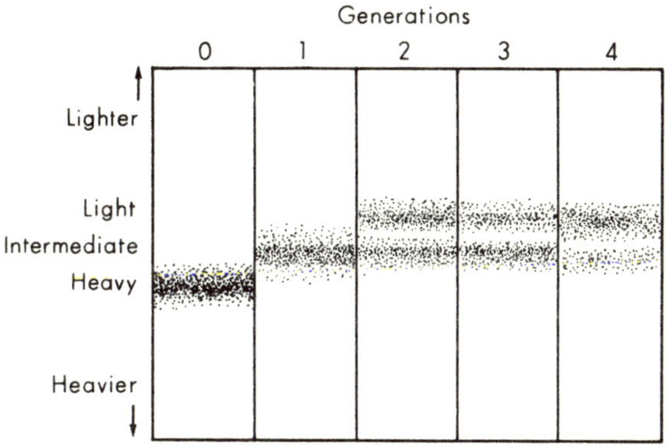

Figure 23. Different density bands of DNA obtained in successive generations during Meselson and Stahl's experiment.

This work was confirmed in 1960 by Noboru Suoeka at Princeton, who used DNA in dividing cells of *Chlamydomonas*, a single-celled green alga. In 1961 it was still further supported by Edward Harvey Simon of Purdue, who worked with a tissue culture of rapidly dividing human cancer cells.

Thus the molecular basis of the gene was clearly established. The question now was: how does DNA control the rest of the cell outside the nucleus? How does it insure the correct formation of the enzymes and other proteins needed for the cell to operate?

9

The Role of RNA

UNTIL THE END OF THE 1930S, SCIENTISTS BELIEVED THAT DNA was characteristic of animal cells and that RNA was only to be found in plant cells. Their chief sources of these two kinds of nucleic acid were calf thymus glands for DNA and yeast for RNA. However, more exact work later revealed that both types of nucleic acid exist in both plant and animal cells. Therefore, once the genetic function of DNA had been established, the natural question was: what is the function of RNA?

Some clues could be found in work done even before the role of DNA was finally elucidated. Back in the 1940s the Swedish researcher Torbjörn O. Caspersson and a Belgian, Jean Brachet, had succeeded in proving that RNA is a universal component of living cells. By using highly refined cytochemical methods, they had also discovered that RNA is mainly distributed in the body of the cell outside the nucleus but is also contained within the nucleus in special pockets called nucleoli. This finding alone was thought-provoking: could it be that

RNA formed some kind of link between the nucleus and the cytoplasm in the rest of the cell?

Caspersson and Brachet went further. They discovered the important—and unexpected—fact that those cells which specialize in making protein are also those which are richest in RNA. Thus cells in organs such as the heart or the kidney, where very little protein is synthesized, are poor in RNA. On the other hand, large quantities of RNA are present in the cells of protein-making organs such as the pancreas or the liver, and especially in the silk glands of silkworms—glands that have the sole function of producing the proteins in silk. It looked as though the synthesis of protein somehow depended on the presence of RNA.

LOCATION OF RNA IN CYTOPLASM

To discover where RNA is distributed in the body of the cell, techniques developed by Albert Claude of the Rockefeller Institute were particularly useful.

Cell scientists had been able to discern various structures within the cell by using ordinary light microscopes, but to investigate them more thoroughly they needed to isolate each kind of structure without destroying its ability to function. Claude's techniques enabled them to do just this.

In 1943 Claude discovered that a tissue such as liver can be broken down by being put into a suitable solution and then disintegrated in a Waring Blendor. This device has a rapidly rotating steel blade that tears cells apart and liberates their contents. By centrifuging the resulting suspension at different speeds, Claude was able to sediment out cell structures of different weights.

For example, by centrifuging at 1,500 times the force of gravity for three minutes, he got the heaviest components—the cell nuclei–sedimenting out. By pouring off the liquid above this sediment, and centrifuging it at 2,000 times the force of gravity for twenty-five minutes, he got smaller cell structures, including the mitochondria. These are relatively large sausage-shaped bodies in which nearly all the energy production of the cell goes on: they are its powerhouses. Finally, by centrifuging at 18,000 times the force of gravity for ninety minutes he obtained small granules, originally known as microsomes. Even then, scientists believed that the remaining solution contained still smaller particles which could not be sedimented out.

Intensive investigations on these microsomes were then launched by George Emil Palade of the Rockefeller Institute and Philip Siekevitz of Harvard. Palade used a then new tool, the electron microscope, which enabled him to peer at structures far too small to be seen with the conventional light microscope. Siekevitz used the techniques of biochemistry. Together, the two men achieved a strikingly successful collaboration.

Chemical analysis showed that the microsomes were a mixture of RNA and protein, containing as much as 40 to 50 per cent of RNA. Further, the RNA in these granules accounted for most of the RNA in the cell.

WHERE IS PROTEIN MADE?

Once the site of most of the cell's RNA had been located, the next step was to find out where protein is made in the cell. In this work, amino acids labeled with radioactive carbon atoms were used to trace where amino acids concentrate in the cell during protein synthesis. Henry Borsook of Cal Tech and Tore J. M. Hultin of the Wenner-Gren Institute in Stockholm independently tackled the problem by this method in 1950.

The procedure they used was to inject radioactive amino acid into a laboratory animal and kill it soon after injection.

After such a short time, the researchers decided, only those proteins being formed would contain much radioactive amino acid. By centrifuging to separate the different cell components of the liver—nuclei, mitochondria, microsomes, and remaining fluid—they could find out which component had the most radioactivity. This would indicate the site of protein synthesis.

They found that the microsomes had more than twice as much radioactivity per gram of protein as the other structures. It seemed likely, therefore, that these were the places where proteins were put together.

What were the microsomes? The electron microscope revealed to Palade that they were small baglike structures bearing many tiny particles in the walls. Previous studies of sections of whole cells with the electron microscope had shown just such tiny particles, attached not to baglike structures but to one side of a folded membrane packed into the body of the cell and called the endoplasmic reticulum. With imaginative insight, Palade guessed that the baglike structures were really fractions of this reticulum, torn apart in the separation process, with edges which had later closed together, thus giving the appearance of little spheres.

This hunch was later proved correct. By using a detergent of bile salt, the membrane was freed from the tiny particles and these could then be collected in a sediment by centrifuging at 100,000 times the force of gravity for two hours. About 55 per cent of the weight of these particles is RNA, the remainder being protein. Because of their high RNA content, the particles are called ribosomes.

A further check on the site of protein synthesis was now made. Borsook and Hultin had shown that proteins were made at the site represented by microsomes, but did this mean that they were built up on the endoplasmic reticulum or on the ribosomes?

To answer this question, Paul Charles Zamecnik of Harvard University, and his co-workers, did experiments in 1956 which followed the pattern of the earlier work by Borsook, except that the actual ribosomes were now separated and checked for radioactivity.

By killing rats at different times after injection with labeled amino acid, they were able to follow the progress of this labeled material in the liver cells. They found that it was taken up very quickly in the ribosomes and later appeared in the protein of the cell fluid. From this they concluded that the cell proteins were synthesized at the ribosomes and moved out into the cell afterward.

Meanwhile, other work on the synthesis of protein had been done, not by using cells in a living animal, but by employing the necessary cell components mixed together in the test tube. Thus in 1953 Vincent George Allfrey and Alfred E. Mirsky of the Rockefeller Institute, together with Marie Maynard Daly, found that protein synthesis going on in cell fragments could be stopped by adding ribonuclease, the enzyme which specifically breaks down RNA. Here was more evidence of the essential role of RNA in protein synthesis.

In 1954 Ernest Frederick Gale and a colleague at the University of Cambridge in England bombarded bacterial cells with high-frequency sound waves and obtained particles which could still synthesize a particular enzyme. To do this, the particles required the amino acids constituting the enzyme, plus a supply of ATP as a source of energy. They also found that adding ribonuclease halted synthesis of the enzyme. Further, if they added more RNA they made enzyme synthesis start up again.

All of this work demonstrated conclusively that RNA at the site of the ribosomes was essential for making protein. What it did not explain was how this RNA operated. Nor did it assign any role to the rest of the RNA present in the cell.

DISCOVERY OF TRANSFER RNA

The first clarification of this confusing picture came from research by Mahlon Bush Hoagland at Harvard. In 1955 Hoagland was studying the synthesis of protein from amino acids in the test tube, using suitable cellular components which had first been separated from each other by centrifuging.

When Hoagland had put together a test-tube system, including ribosomes, that would make protein it occurred to him to wonder whether the same system would build up RNA from nucleic acid precursors—a not dissimilar operation.

He therefore set up two parallel experiments, using the same basic system of cell components in each. To one, he added only radioactively labeled amino acids; to the other, only radioactively labeled precursors of RNA. He found that, as he had hoped, these precursors turned up in RNA, which was therefore radioactively labeled.

But Hoagland also found, to his complete surprise, that the test-tube system to which he had added radioactive amino acids now contained radioactive RNA too! Somehow, the radioactivity in the amino acids had gotten mixed up with the ordinary ribosomal RNA in the test-tube system. But why?

By applying chemical techniques, Hoagland found that he could prize the radioactive amino acid away from the RNA. After this, the RNA was left as before and was not radioactive.

However, this RNA turned out to be different from ribosomal RNA. Even if whirled in the ultracentrifuge at 100,000 times the force of gravity, it would not separate out. This meant that it must be a relatively small molecule, much smaller than ribosomal RNA, and soluble in the cell fluid. It was christened soluble RNA for this reason.

Significantly, Francis Crick and a Belgian biochemist, H. Chantrenne, had already predicted that, in protein synthesis, amino acids might be carried into place on an assembly line by some kind of adaptor molecule. Such an adaptor molecule would have to join up loosely with an amino acid before towing it into place on the assembly line. Could soluble RNA be this adaptor?

Further research revealed that there is not just one kind of soluble RNA but over twenty different kinds, each specific for one kind of amino acid. This was proved when experiments were carried out to load soluble RNA with radioactive amino acid. It was found that eventually the soluble RNA became saturated with one kind of amino acid and would not take up any more. However, if another kind of amino acid were then added to the mixture, the soluble RNA would absorb it. In other words, more than one kind of soluble RNA was present. By extension, this technique led to the identification of a soluble RNA for each amino acid.

In 1957 several laboratories reported that soluble RNA transfers labeled amino acids to ribosomes. Following Crick's earlier speculation, it now seemed likely that soluble RNA, or transfer RNA as it is often called, lined up the amino acids on the ribosome in such a way that they fitted into the right sequence to form a particular protein.

STRUCTURE OF TRANSFER RNA

From physical and chemical studies we now know that transfer RNA contains only a single nucleotide chain of about eighty nucleotides. It seems to be formed by transcription from a section of DNA.

The complete nucleotide sequence of a transfer RNA molecule was first worked out in 1965 by a team of researchers at Cornell University under Robert W. Holley. This was the transfer RNA which is specific for alanine and it turned out to contain seventy-seven nucleotides, several of them being quite unusual.

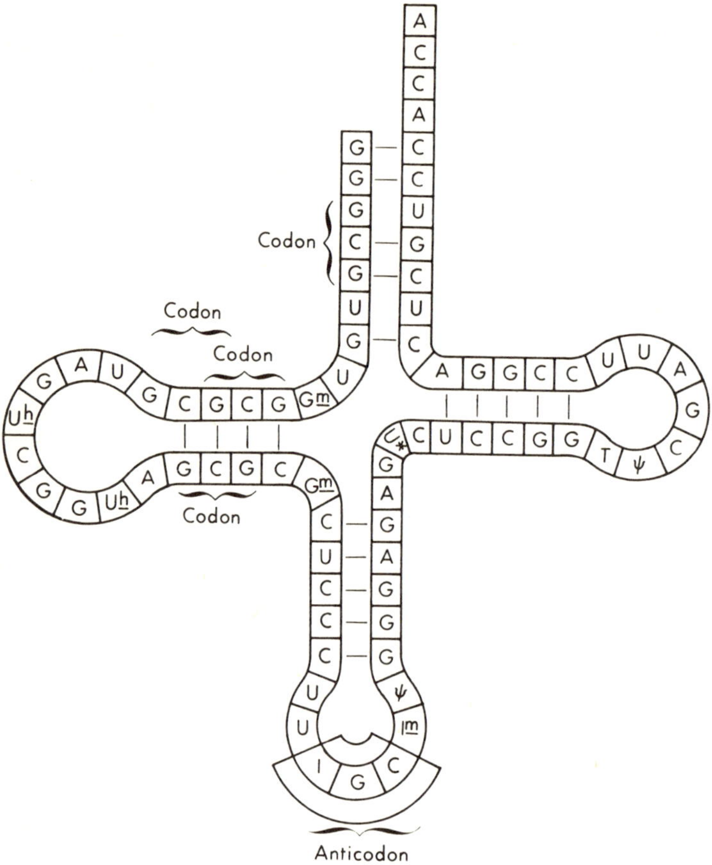

Figure 24. One model proposed by Holley for the transfer RNA for alanine from yeast. A = adenine, C = cytosine, G = guanine, and U = uracil. Among the unusual components is inosine (I), a derivative of adenine. If C pairs with G and A with U, short double-strand helical regions will form. The IGC group about halfway along the molecule may represent the anticodon (see Chapter 10) for attachment to messenger RNA. Alanine may be attached at one of the GCG codons in the molecule.

Using transfer RNA obtained from yeast, the researchers split it into nucleotide fragments of different kinds by means of two varieties of ribonuclease enzyme. These fragments were then separated and identified with the help of chromatographic techniques. After this, the researchers had to deduce how the pieces had been fitted together, just as Stein and Moore worked out the sequence of amino acids in ribonuclease.

The physical structure of transfer RNA had already been investigated by Maurice Wilkins and his colleagues at King's College, London, using material crystallized from yeast. In 1962 this group reported that the diffraction patterns indicated that this crystallizable RNA formed a two-stranded helix. Because other data indicated that transfer RNA is single-stranded, the researchers concluded that transfer RNA must be a single strand folded back upon itself. Two possible configurations were suggested: a hairpin-like shape with both ends of the chain at one end of the helix, or a helix broken at one side.

However, this idea was short-lived. Two years later, Michael Spencer, who had been one of the group, announced that less than 20 per cent of the crystallizable RNA was transfer RNA, the remainder being of ribosomal origin. The hairpin model was therefore incorrect, although some helical regions could still exist.

The analysis of transfer RNA for alanine by Holley fits in with this idea, because he has found short sequences of nucleotides that can pair. So far, however, the exact configuration of any transfer RNA molecule is unknown.

RNA THAT IS A COPY OF DNA

Although the discovery of transfer RNA was highly important, it did not solve the problem of how DNA controls the formation of protein. Scientists reasoned that some complementary pattern, or template, must exist to insure that transfer RNAs brought their amino acids into line in the correct sequence for making any particular protein. This template would of course embody the genetic instructions embodied in the DNA in the nucleus. Further, since they knew that protein synthesis occurs on the ribosomes, researchers supposed that the template must be there, in the ribosomal RNA.

As time went on, however, this hypothesis seemed more and more inadequate—for two reasons. First, it implied that each ribosome would be suitable for synthesizing only one particular protein, but if this were so it was almost impossible to account statistically for the flexibility cells display in the way in which they can switch from making a lot of one protein to making a lot of another.

Second, the ratios of the four bases—adenine, guanine, cytosine, and uracil—on the RNA of the ribosomes bore no relation to the ratios of the equivalent bases in the DNA from the same species of organisms. Without this relation, the ratios could hardly represent a copy of the information in the DNA.

A way out of this dilemma was first hinted at in an observation of Hershey's. After his work with Chase in 1952 which proved that the hereditary material of bacteriophage is DNA, Hershey was studying how new phage DNA is synthesized in a bacterium which has been infected by the phage.

As before, Hershey was using radioactively labeled phosphate as a tracer. He noted in passing that there seemed to be a small proportion of bacterial RNA which rapidly became radioactive after labeled phosphate was added to the culture medium. Apparently the infected bacterium was making new RNA as well as new phage DNA. Why was this?

Hershey could not answer this question; and three years later, in 1956, Elliot Volkin and Lazarus Astrachan of the Oak Ridge National Laboratory followed up the line of inquiry. They discovered that not only was new RNA synthesized in the bacterium after infection but—still more important—the makeup of this RNA mimicked that of the DNA in the invading phage!

Evidently the phage had succeeded in disrupting normal RNA production in the bacterium and had imposed a new pattern in which phage-tailored RNA was synthesized. Here, then, was evidence that some RNA did exist which mimicked DNA, and that such RNA was involved in the synthesis of new phages.

After this, several workers showed conclusively that a type of RNA can be synthesized which depends directly on the presence of DNA and is a copy of that DNA.

One group, Jerard Hurwitz and John J. Furth at the New York University School of Medicine, stumbled upon the mechanism of this synthesis by chance. Their work followed that done in 1955 by Marianne Grunberg-Manago and Severo Ochoa at the same school, in which they discovered that polynucleotide phosphorylase could link together nucleotides with two phosphate groups and make synthetic RNA.

Hurwitz and Furth suspected that this enzyme was not the one used in the living cell to make RNA. They believed RNA would be assembled from more energy-rich nucleotides, having three phosphate groups. Accordingly, they began to hunt for another RNA-forming enzyme.

Using *Escherichia coli* as a source, they eventually obtained a cell-free extract which, in the presence of a primer of RNA from the *E. coli*, would link up triphosphate nucleotides to make more bacterial RNA. On purifying the extract, they managed to obtain a single enzyme which was doing this work. They christened it RNA polymerase.

They also purified the RNA which they were using as a primer and then they had a shock: they found that the RNA had been an effective primer only because it had been contaminated with DNA! In fact, it was this bacterial DNA which stimulated the synthesis of bacterial-style RNA.

Now Hurwitz and Furth checked the proportions of bases in the DNA which they had unwittingly used as a primer and compared these with the proportions of bases in the synthetic bacterial-style RNA. They found that the synthetic RNA mimicked the makeup of the primer DNA.

Pursuing their discovery still further, the New York researchers tried using a special synthetic DNA as a primer. They employed two synthetic DNA polymers, one containing only thymine and the other containing alternate units of thymine and adenine. How would these synthetic primers affect the RNA being formed?

Hurwitz and Furth found that when they used DNA made of thymine, the RNA was wholly composed of adenine, the regular complement to thymine. When they used DNA made of thymine and adenine, the RNA that resulted contained a matching sequence of uracil and adenine. Even though all four nucleotide bases were available in the test tube, only those complementary to the DNA were incorporated in the new synthetic RNA.

Concurrently with this research, which was reported in 1959, Samuel Bernard Weiss at the University of Chicago and Audrey Stevens at St. Louis University demonstrated the same reliance of a synthetic RNA upon the sequence of bases in DNA.

Further work on the problem was done by Sydney Brenner of Cambridge, François Jacob, a geneticist from the Pasteur Institute in Paris, and Matthew S. Meselson working together at Cal Tech in 1961.

They began by growing *E. coli* bacteria in a culture medium containing heavy isotopes of carbon and nitrogen, C-13 and N-15. As a result, the bacteria made ribosomes which were heavier than normal. Then the researchers infected the bacteria with bacteriophage and immediately transferred them to a culture medium containing ordinary carbon and nitrogen. Any ribosome synthesized after infection would therefore be made of protein and RNA containing ordinary carbon and nitrogen, and so be lighter than the ribosomes previously synthesized which contained C-13 and N-15.

Besides labeling the ribosomes in this way, the researchers exposed the bacteria to radioactive phosphate while infecting them with the phage. This meant that any phage-style RNA made by the bacteria after infection would be tagged with radioactive phosphorus.

After a few minutes of incubation, the infected bacteria were broken open and their contents were centrifuged in cesium chloride solution. The fractionally heavier ribosomes containing C-13 and N-15 formed a lower band in this solution than the ordinary ribosomes did.

Now the researchers checked where the newly formed phage-style RNA was, using the radioactive phosphorus as their guide. They found that this new RNA was in the same band as the old ribosomes formed before infection.

In a parallel experiment, the Cal Tech team labeled the protein formed after infection, instead of labeling the RNA. To do this, they used radioactive sulfur, which was incorporated in the amino acids methionine and cysteine. They found that the newly made protein was also associated with the old ribosomes, along with the newly formed RNA.

Clearly, synthesis of phage protein could proceed on the ordinary bacterial ribosomes. As these ribosomes did not contain any phage RNA, they obviously could not embody genetic instructions for making phage protein. These instructions were available only from the new RNA synthesized after infection. What the Cal Tech investigators had found indicated that the ribosomes themselves were merely non-specific protein assembly plants, which could be "hired out" to any RNA requiring them.

MESSENGER RNA

Shortly after, a suggestion was put forward which brilliantly knit together all of these observations and framed an explanation for them. Its authors were François Jacob and Jacques Monod, a biochemist at the Pasteur Institute in Paris.

Instead of trying to visualize the RNA of the ribosomes, where the proteins are assembled, as a repository of genetic information, they boldly proposed that the DNA-like RNA which had been discovered had the specific role of carrying genetic instructions from the cell nucleus to the ribosomes. They suggested that this special RNA be called messenger RNA.

The function of the RNA already in the ribosomes was not, of course, explained by this hypothesis. However, it did at least offer a working model for the all-important transmission of information from DNA to the protein assembly plants.

In the same year, further work was reported by a group of workers including James D. Watson at Harvard and François Gros of the Pasteur Institute. This group had traced the fate of messenger RNA which was synthesized in ordinary, uninfected bacteria.

To do this they tagged the RNA with heavy carbon, which they provided by growing the bacteria in a medium containing uracil that incorporated heavy carbon. They then took samples of the bacterial contents at different times after supplying the tagged uracil, and centrifuged these contents to separate the constituents.

They found that the heavy carbon appeared in a few minutes in messenger RNA but then left this equally quickly to appear in ribosomal RNA instead. Evidently the messenger

RNA had a very short life on its own and was then incorporated into the ribosomes.

The picture emerging from this research was of messenger RNA that was being continually sent out from the genes to the ribosomes, where it was used and then replaced by fresh messenger RNA. In 1963 Cyrus Levinthal and his colleagues at the Massachusetts Institute of Technology showed that in another bacterium messenger RNA is used approximately fifteen times for the synthesis of the corresponding proteins before it is discarded. In mammalian cells, however, the turnover of messenger RNA is not nearly so rapid.

TRANSCRIPTION OF DNA

For the genetic message to be transcribed from DNA to messenger RNA by some base-pairing procedure, it is pretty obvious that the first step must be for the twin nucleotide strands in the DNA molecule to separate. Only when the bases in each strand are exposed can a complementary strand of messenger RNA be built up on them.

When the existence of this DNA-like RNA was first being investigated in test-tube reactions, researchers believed that this RNA mimicked the total DNA molecule. In other words, two RNA strands were made, one from each DNA strand, and then the two RNA strands combined to make a double helix.

However, it seems that this is not the way things happen in the living cell. A check on Volkin and Astrachan's work on phage RNA was made in 1962 by Ekkehard K. F. Bautz and Benjamin Downs Hall at the University of Illinois. They found that although the amount of guanine and cytosine in the phage messenger RNA was about the same percentage of the total nucleotide content as it was in the phage DNA, the individual bases were in different proportions in the messenger RNA. This meant that the guanine and cytosine could not pair off equally with each other—and hence could not occur in a double helix. The alternative was to suppose that RNA is a single-stranded molecule.

If messenger RNA is a single-stranded molecule, it follows that it is transcribed from only one strand in the DNA molecule. That this is in fact true has been elegantly demonstrated by M. Hayashi, M. N. Hayashi, and Sol Spiegelman of the University of Illinois, who announced their results in 1963.

In their investigation, this group used a bacteriophage designated phi-ex 174 which, in its resting state, contains single-stranded DNA. When the phage invades a bacterial cell, this single-stranded DNA forms a second, complementary strand upon itself and so becomes the usual double helix.

Synthesis of phage-style messenger RNA begins only after the phage DNA has become double-stranded. Fifty minutes after infection, the amount of messenger RNA present is substantial. At this point the Illinois researchers isolated this messenger RNA, which was labeled with heavy hydrogen (tritium), by using column chromatography of the kind used by Stein and Moore in separating amino acids from each other.

Having isolated the messenger RNA, the Illinois group found that it formed a double helix with one of the strands from the double-stranded DNA molecule which the virus formed in the infected bacterium. This showed that the messenger RNA was single-stranded. Further, it was a strand which complemented one of the DNA strands.

The DNA strand which was the complement to the messenger RNA was obviously the strand from which the messenger RNA was transcribed in the first place. Which of the two DNA strands was used for the transcription?

The three researchers found that the messenger RNA

would not form a double helix with the single-stranded DNA found in the resting stage of the phage. Evidently, therefore, it had been transcribed from the complementary DNA strand formed after infection.

This conclusion was supported by an analysis of the base composition of the messenger RNA and the single-stranded resting form of the DNA. The proportions of the four bases turned out to be identical in the two molecules. Hence the RNA must have been transcribed from the complement to the resting form of the DNA.

Other experiments on different organisms have confirmed this general finding. We can therefore imagine that the enzyme responsible for building up messenger RNA, namely RNA polymerase, splits the two chains of a DNA molecule apart like someone undoing a zipper, thus allowing the RNA to build up along the exposed bases of one chain. This means that one DNA strand embodies the genetic message that is actually used, while the other strand acts merely as a "keeper."

This gives us a picture of the transcription of genetic information from DNA to messenger RNA, but it does not tell us anything about the way in which this information is embodied in the nucleic acid.

This genetic information must govern the way in which amino acids are assembled at the ribosomes into individual proteins. The essential character of nucleic acid is that it consists of long molecules composed of four different nucleotides strung together in varying sequence. Scientists therefore supposed that the nature of the sequence was itself the message transcribed from DNA to messenger RNA and then carried to the ribosomes. According to this view, the message would be "read" at the ribosomes when the various transfer RNAs, each bearing its own particular brand of amino acid, linked up with the messenger RNA in the order dictated by the sequence of the bases in its nucleotides. The amino acids would then join together to form the polypeptide chains of individual proteins, and the translation of the genetic message into protein would be complete.

This view assigned to the sequence of nucleotide bases the role of a code message. Just as two kinds of symbol, a dot and a dash, can be grouped in various ways in the Morse Code to spell out ordinary words, so the four nucleotide bases were thought of as symbols forming different groupings, each of which "spelled out" a single amino acid. Comparing a protein molecule with a sentence, we can regard each amino acid as a word within that sentence. In the genetic code, we then have a code word "spelled" in nucleotide bases which stands for the amino acid word. (Indeed, we have two code words: the word in the DNA molecule, and its complement as transcribed in the messenger RNA molecule.) But what is the code, and how is it "read" at the ribosomes? Molecular biologists everywhere were eager to know the answer to these crucial questions. Understanding the nature of the genetic code was now the central problem of research on DNA.

10

The Genetic Code

IN TRYING TO UNDERSTAND THE GENETIC CODE, SCIENTISTS FACED the basic difficulty that, although they knew the amino acid sequence of various proteins, they did not know the corresponding sequence of bases in DNA. However, by speculation alone it seemed clear that at least three bases must be needed to specify each amino acid.

The reason for this had been stated as far back as 1952 by Alexander Latham Dounce, a biochemist. Two years later it had been enlarged upon by George Gamow, a theoretical physicist now at the University of Colorado.

Suppose only one base were to be used as a code word for one amino acid. Evidently, the four bases would suffice to specify only four amino acids. As there are twenty-odd amino acids, this is inadequate.

What if two bases are used for each code word? The four bases can be arranged in sixteen possible pairs: this is still not enough. Hence we know that at least three bases are required. These yield a total of sixty-four possible combinations, which is more than adequate.

AAA	AAG	AAC	AAU
AGA	AGG	AGC	AGU
ACA	ACG	ACC	ACU
AUA	AUG	AUC	AUU
GAA	GAG	GAC	GAU
GGA	GGG	GGC	GGU
GCA	GCG	GCC	GCU
GUA	GUG	GUC	GUU
CAA	CAG	CAC	CAU
CGA	CGG	CGC	CGU
CCA	CCG	CCC	CCU
CUA	CUG	CUC	CUU
UAA	UAG	UAC	UAU
UGA	UGG	UGC	UGU
UCA	UCG	UCC	UCU
UUA	UUG	UUC	UUU

That there is such an abundance of possible combinations suggests that some of them mean the same thing since only twenty amino acids are specified by the code. If so, the code must be termed, in coding jargon, "degenerate."

Further, some combinations could specify amino acids while others might have some other function—say, to act as some form of "punctuation," such as a period, or a capital letter to mark the beginning of a message.

IS THE CODE OVERLAPPING?

Granted that the genetic code must be a triplet one, at least, another basic question which arises is whether the code is read at the ribosomes in straight sequence or whether it is overlapping.

In straight sequence, the first three bases would indicate one amino acid, the second three another amino acid, and so on. Alternatively, bases 1, 2, and 3 might code for one amino acid, and bases 2, 3, and 4 or bases 3, 4, and 5 for another amino acid–in which case the code would be an overlapping one.

Even without the knowledge of what amino acids are specified by particular triplets of bases, it proved possible to devise experiments to test the basic nature of the genetic code.

If the code were overlapping, it would cut down on the freedom with which sequences of bases could be arranged. As a result, certain amino acids would frequently appear together in proteins. However, this does not occur, and for this reason alone scientists considered that an overlapping code was unlikely.

An experiment to decide whether or not the code is overlapping was performed by a German worker, H. G. Wittmann of Tübingen. In an overlapping code, a change in one base should, as a rule, result in alteration to two or three adjacent amino acids which have been joined together as a result of the code instructions. To determine the effect of altering bases, Wittmann treated the tobacco mosaic virus with nitrous acid, which changed some of the nucleotides in its genetic material. (In this virus the genetic material is RNA, not DNA.) The virus was then used to infect tobacco plants and so produce new viruses in the leaves according to the instructions of the modified genetic material. What was the effect on the newly made virus protein?

This particular protein is rather simple, consisting of a single chain of 158 amino acids, and the precise sequence of these had already been worked out. Wittmann now checked whether this sequence had been modified. He found that although changes had occurred, they usually involved only one amino acid and in no case were two or three adjacent amino acids altered. From this it followed that the genetic code could not be overlapping.

IS THERE A FIXED STARTING POINT?

The next problem was to determine how the code is read without confusion as to the starting point. Thus, suppose there is a sequence in the middle of a gene reading

C A T C A T C A T C A T.

What determines whether this is read as

CAT CAT CAT CAT

or as

C ATC ATC ATC AT?

We have the same problem in reading print. If you did not know that you have to begin at the left-hand margin of each line, you would be unable to decipher the message of the words on this page.

In 1961, highly ingenious experiments were carried out at the University of Cambridge in England by Francis Crick and his colleagues Mrs. Leslie Barnett, Sydney Brenner, Richard J. Watts-Tobin, and Robert Shulman to determine whether the genetic code is read from a fixed starting point.

The Cambridge workers used the T4 virus, a bacteriophage which infects *Escherichia coli* bacteria. The genetic sequence in this virus had already been mapped with astonishing precision by Seymour Benzer of Purdue University, using techniques which stemmed from those pioneered by Morgan when he mapped the chromosomes of the fruit fly, *Drosophila*. The Cambridge workers now wanted to determine the genetic effect of altering a single base in the DNA of the virus.

The basic procedure in their experiments was the genetic cross. If two different varieties of the T4 virus infect a bacterial cell, the resulting new viruses will be of three kinds: two as before, plus a third which has some characteristics from both the others. This third kind occurs because some of the DNA in the first two viruses splits up, and then these fragments join up with each other during the formation of new viruses.

If one T4 virus has a defect at point A and another has a defect at point B in the same gene, a cross between the two viruses will yield some progeny having both defects. How often this happens depends how close A and B are to each other. The closer they are, the less likely it is that both defects will appear in one virus.

Benzer used these facts to map genes in the T4 virus. He demonstrated that each gene has many different sites where changes can occur, and that these sites are arranged in line.

Crick and his colleagues used a substance called proflavin to induce mutations in the T4 genes. They had evidence that proflavin operates by adding or deleting a base in the DNA of the virus, rather than by changing one.

Now they crossed viruses in which a base had been added or deleted in this way. Usually, the newly formed viruses that resulted were mutants too. Occasionally, however—perhaps once in a million times—a normal virus turned up. This meant that the affected gene was functioning again. Somehow, the action of the proflavin had been overruled. But how?

To find out, the Cambridge team investigated the detailed structure of such an apparently normal gene, using Benzer's technique of fine genetic mapping.

One way in which the gene might function normally again would be if the altered base had been corrected by some means. However, Crick and his co-workers found that the usual reason was quite different: there was a further addition or deletion at a nearby site along the gene. This further change, of itself, could also remove the function of the gene, yet when it was present along with the first change it allowed the gene to function normally (provided it was not too far away from the first change).

Crick explained this by assuming that the triplet code was read in sequence from a definite starting point. In that event, addition of a base would alter the sequence of all triplets following it—and so destroy the functioning of the gene. If a second defect were then caused by removal of a base, the sequence after that point would revert to normal. Although the triplets between the two changes would not make sense, the rest of the message would be restored and so the gene would function fairly well.

Original gene	C A T/C A T/C A T/C A T/C A T
Base added	C A T/G C A/T C A/T C A/T C A
Base added plus base removed	C A T/G C A/C A T/C A T/C A T

For this kind of correction to work, the defects must be of opposite kinds. The Cambridge workers designated these plus and minus. Whether a "plus" defect represented the addition of a base or the subtraction of a base was immaterial: the point was that its effect was contrary to that of the "minus" defect.

Crick now went a step further and produced mutants containing three defects of a like nature—that is, three plus defects or three minus defects. What was the result? Such a triple mutant was functional!

"This is clearly a remarkable result," commented Crick later. "Two blacks will not make a white but three will."

Here was hard evidence for the supposition that the genetic code works in triplets. The three alterations would then have the effect of an intruder triplet, beyond which a normal sequence would be resumed.

READING THE CODE AT THE RIBOSOMES

Before this work by the Cambridge team, it had already been shown that proteins are assembled in an orderly linear sequence at the ribosomes. This work was performed by Richard Schweet of the University of Kentucky Medical School and Howard Marvin Dintzis at the Massachusetts Institute of Technology.

These two researchers added radioactively labeled amino acid to a system of ribosomes synthesizing hemoglobin, the protein in red blood cells which is responsible for the carriage of oxygen. They then analyzed the position of this amino acid in the hemoglobin at different times after adding it to the mixture.

If the labeled amino acid were distributed at any of its possible sites along the hemoglobin chain, the researchers would know that it was taken up in no particular sequence. If, on the other hand, the amino acids were taken up in orderly sequence, the labeled acid would be grouped along its possible sites at one end of the chain.

What Schweet and Dintzis discovered was that the synthesis of the hemoglobin protein went in orderly sequence. It began with the amino acid valine at one end and went on, bond after bond, to the other end. New amino acids were added to the growing chain at the rate of about two a second, the researchers calculated. This meant that the complete molecule of 150 amino acids was built up in 1½ minutes.

How is this orderly reading of the code achieved? Alexander Rich and his associates at the Massachusetts Institute of Technology have provided part of the answer from their studies of protein synthesis in rabbit blood cells. In 1962 they discovered that ribosomes cluster together during protein synthesis, often in fives. They named such clusters polysomes. The ribosomes in these clusters are arranged in line, they found, and the electron microscope reveals that they are linked by very thin threads. Later studies have shown similar groupings in some other mammalian cells, with up to forty ribosomes in a single polysome.

By special staining techniques, Rich succeeded in demonstrating that the very thin threads linking the ribosomes are composed of RNA. It may be, therefore, that this is messenger RNA, with one continuous thread linking different ribosomes.

If this is so, the reading mechanism may involve each ribosome's rolling along the messenger RNA while the transfer RNA molecules attach themselves at the correct places on the messenger molecule as each portion of it is passed over by the ribosome. Proceeding in sequence along the messenger RNA molecule, each ribosome would thus bring together the correct sequence of transfer RNA molecules, each bearing its own amino acid. These amino acids would then be linked in orderly sequence by the action of an enzyme.

MATCHING LINEAR SEQUENCES

If the genetic code is linear, being read from a fixed starting point, and if the assembly of amino acids also proceeds in an orderly linear fashion, it follows that the sequence of amino acids in a protein will exactly match the sequence of controlling sites in the gene. That this is in fact so has been demonstrated by Charles Yanofsky and his associates at Stanford University.

In 1964 these researchers reported that they had analyzed the position of sites causing mutation in one gene of *Escherichia coli.* This gene controls the production of one protein chain in an enzyme, tryptophan synthetase, which is responsible for the assembly of the amino acid tryptophan. Mutations in the gene lead to the substitution of incorrect amino acids in the protein chain.

The Stanford workers found that the relative positions of these incorrect acids along the chain exactly corresponded to the relative positions of the mutant sites along the gene. Here was excellent proof of the linear nature of the genetic code.

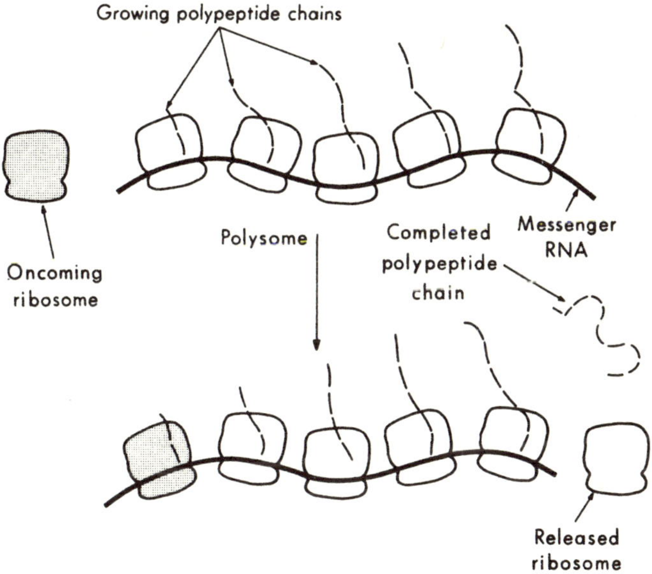

Figure 25. Messenger RNA may be read at a group of ribosomes called a polysome. Each ribosome in turn would "read" the message in the long RNA molecule by binding transfer RNA-amino acid complexes in the correct sites on the messenger RNA so that the amino acids could join up with each other to form a growing polypeptide chain.

DECIPHERING THE CODE WORDS

The first announcement that the code word for an amino acid was known came at the Fifth International Congress of Biochemistry, held in Moscow in 1961. The Congress was held in a vast, gloomy skyscraper belonging to the Lomonosov University and it was so spread out that only a handful of people turned up to hear a paper given by an obscure young biochemist from the National Institutes of Health named Marshall W. Nirenberg. So momentous was his news, however, that Matthew Meselson, who was in the audience, hurried to tell it to Francis Crick, who then invited the young biochemist to repeat his paper to a large, selected audience.

What Nirenberg had done was extraordinarily simple in outline and extraordinarily effective in practice. He set up a system which would synthesize protein in the test tube when messenger RNA was supplied, and he then supplied this system with messenger RNA tailored to his own specification. By examining the resulting protein molecules, he could then deduce what bases in his messenger RNA corresponded to the amino acids in the protein. And, of course, the bases in the messenger RNA in turn indicated what DNA bases corresponded to the amino acids.

To carry out this brilliant maneuver, Nirenberg and his associate J. H. Matthaei made synthetic messenger RNA according to the recipe first devised by Marianne Grunberg-Manago. She had made an RNA-like polymer out of just one base repeated over and over. Severo Ochoa had then demonstrated that all of the four bases could be used in any combination to make such artificial RNA.

Nirenberg and Matthaei began by making poly U, a polymer assembled from only one base, uracil. They then used extracts of *Escherichia coli* bacteria to assemble a test-tube system of ribosomes, transfer RNA, and the necessary enzymes. To this they added radioactively labeled amino acids, plus ATP as a source of energy. Alone, this system did not manufacture protein out of the labeled amino acids. However, when the two researchers added their custom-built poly U, a tremendously rapid build-up of protein occurred.

To detect formation of this protein, they precipitated out all of the protein present and measured its radioactivity. The original protein in the system was not radioactive, but the new protein was because it was formed from the labeled amino acids.

Although they knew that protein was being made from these amino acids, Nirenberg and Matthaei did not know which were used. By trying out different combinations of the twenty-odd amino acids, they eventually found that only one was needed to make protein with the help of poly U, and that one was phenylalanine.

Thus the two researchers had proved that uracil in messenger RNA directs the incorporation of phenylalanine into protein. In so doing, they had "read" one word of the genetic code:

uracil = phenylalanine

Although this was a tremendous breakthrough, it did not of itself indicate how many uracil bases are needed to specify one amino acid. However, by assuming that the genetic code must be a triplet one, Nirenberg and Matthaei felt justified in writing

UUU = phenylalanine

Independently of each other, Nirenberg and Matthaei in Bethesda, Maryland, and Ochoa and his collaborators in New York now set out to synthesize and test polymers containing every possible combination of the four bases.

Six varieties of synthetic RNA can be made from only

two bases. They are poly AC, poly AG, poly AU, poly CG, poly CU, and poly GU. Although the bases probably unite in a random order we can figure out the chance of finding any particular sequence of three bases. For instance, if poly UC contains 70 per cent of uracil and 30 per cent of cytosine, then 34 per cent of the triplets in the polymer should be UUU, and 14.7 per cent should be UUC.

By analyzing the relative amounts of each amino acid incorporated into protein under the guidance of their synthetic RNAs, and comparing them with the relative proportions of different triplets, the two groups of researchers were able to assign triplets to every amino acid.

However, they still did not know how the three bases in each triplet followed each other in sequence. So far, therefore, they could not give the correct spelling of the code word, or codon, for each amino acid, although they knew which letters were used to make up the word.

SPELLING THE CODONS

The first breakthrough in spelling a codon was announced at the Sixth International Congress of Biochemistry in New York in 1964. There, Marshall Nirenberg, along with his colleague Philip Leder, was able to repeat his triumph at the previous congress in Moscow by stating that he had found out how to spell the codon for the amino acid valine.

Nirenberg and Leder's work involved the binding of transfer RNA to the ribosome. The importance of the codon belonging to the messenger RNA as a site for binding the transfer RNA at the ribosome had already been demonstrated two years earlier. At that time, it was unclear whether transfer RNA itself joins on to the messenger RNA at the ribosome or whether its associated amino acid does.

To settle the matter, a delightfully ingenious experiment was carried out by François Chapeville and François Lipmann of the Rockefeller Institute, along with Günter von Ehrenstein at Johns Hopkins University and Seymour Benzer, Bernard Weisblum, and William Jackson Ray, Jr., of Purdue University.

This experiment involved linking an amino acid to its specific transfer RNA and then altering it before the transfer became positioned on the ribosome. If the transfer RNA still joined up with the messenger RNA at the ribosome, the researchers argued, it would prove that the transfer RNA, not the amino acid, was the binding agent.

The amino acid chosen for the experiment was radioactively labeled cysteine. This was linked to its specific transfer RNA and then altered by treatment with a special catalyst called Raney nickel, which is a suspension of the metal. This catalyst removed the characteristic sulfur from the cysteine, so converting it into another amino acid, alanine.

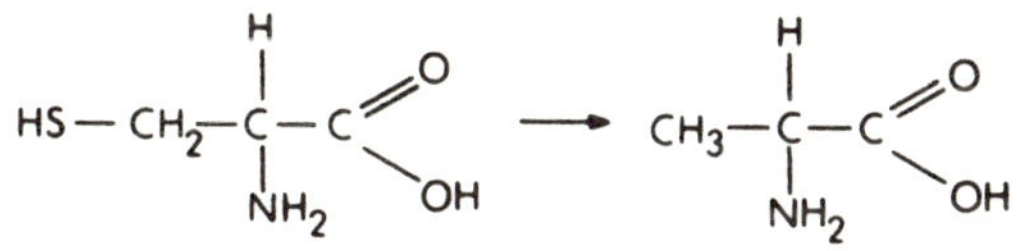

cysteine alanine

Nirenberg and his colleagues had already shown that transfer RNA carrying cysteine will bind on to a synthetic messenger RNA composed of random sequences of uracil and guanine. They had also been able to figure out that the codon for cysteine contained two Us and one G, but they did not know the order of these bases. However, they did know that this combination of bases does not code for alanine.

Chapeville, Lipmann, and their co-workers therefore offered their cysteine-specific transfer RNA, still carrying alanine, the chance of binding on to the synthetic messenger

RNA for cysteine. They found that it did indeed bind, and the labeled alanines joined together to form a radioactive polypeptide. This proved that the transfer RNA, not the amino acid, responds to the code embodied in the messenger RNA molecule.

How would the transfer RNA respond to the code? Surely by a set of complementary bases of its own, which could link up with the appropriate ones in the messenger RNA. Such a codon-complementing set of bases is sometimes called an anticodon, or even a nodoc! In keeping with this idea, it is encouraging to find that the analysis Robert Holley has

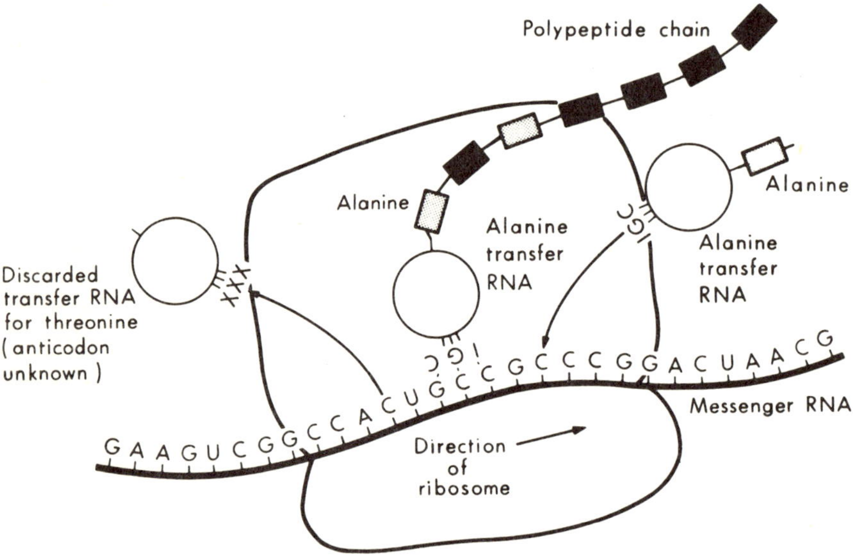

Figure 26. Transfer RNA "reads" the instructions in messenger RNA at the ribosome by means of an anticodon. This pairs briefly with the codon on the messenger RNA that stands for the particular amino acid which the transfer RNA is carrying. When the amino acid has been joined to the growing polypeptide chain, the transfer RNA is released. During chain growth, the ribosome travels along the messenger RNA, allowing reading of one codon after another.

done on the transfer RNA for alanine reveals at least one suitable anticodon in the sequence of bases.

Nirenberg and Leder found that fragments of messenger RNA, instead of whole molecules, would settle on ribosomes and bind transfer RNA to them at the ribosome surface. They then hit on the brilliant idea of synthesizing three-base fragments, containing bases in known order, in order to find which would bind the transfer RNA for any particular amino acid.

To make these triplets, Nirenberg and Leder broke up an artificial polymer of uracil and guanine assembled in random order, using chemicals and enzymes to do so. All sorts of fragments were produced by this method, including two-base as well as three-base units. However, they found that a minimum of three bases was needed to stimulate the binding of transfer RNA.

By painstaking use of paper chromatography and electrophoresis, the two researchers were able to separate the three-base units, which were GUU, UGU, and UUG. They then separated these from each other by further chromatography.

Now Nirenberg and Leder prepared twenty different batches of transfer RNA and added to each a different radioactively labeled amino acid. They also added ribosomes and one or other of their three-base fragments. Where a fragment coincided in spelling with the requirements of a particular transfer RNA, that transfer RNA and its associated amino acid were bound to the fragment at the ribosomes. To detect this binding, the two researchers washed the ribosomes through special filters which allowed ordinary ribosomes through but held back those which carried transfer RNA and amino acid molecules and so were larger. The radioactivity of this deposit, plus a knowledge of which of the twenty main batches had been washed through the filter, told the researchers which labeled amino acid had been bound to the ribosomes.

In this way they discovered that, out of their three synthetic codons, GUU, UGU, and UUG, only GUU stimulated binding, and it stimulated binding of only one amino acid, valine. Here, then, was the first correct "spelling" of a word in the genetic code of messenger RNA:

GUU = valine

Nirenberg quickly followed up this initial success by testing other synthetic codons. Soon he had the correct spelling for several other amino acids.

Besides this, he proposed an explanation for the degeneracy of the genetic code. He suggested that codons with identical sequences in the first and second positions served as codons for the same amino acid.

This method of synthesizing codons and testing whether they stimulate the binding of a particular transfer RNA to the ribosome was enthusiastically taken up by Har Gobind Khorana and his associates at the University of Wisconsin. From their work, still more codons have been elucidated.

Khorana's group has also substantiated the whole idea of a nonoverlapping triplet code. They did this by making studies of synthetic messenger RNAs in which the order of bases was exactly known. To synthesize such messenger RNAs was a very considerable feat and a great advance on the linking of a random sequence of bases which Ochoa and also Nirenberg had used before.

Two general approaches were used: to build up a DNA molecule by adding one nucleotide at a time through chemical procedures; or to make a two-nucleotide molecule and then polymerize it. Either of these was then used as a template for RNA polymerase enzyme to assemble matching RNA molecules out of a suitable supply of nucleotides.

Having made synthetic messenger RNAs to exact specification in this way, Khorana and his co-workers studied what kinds of polypeptides were assembled on them. They found that synthetic messenger RNA which read UCUCUCUCUCU

. . . led to the synthesis of two amino acids, serine and leucine, in alternating sequence. With a triplet code, this could be explained by the fact that the RNA sequence could equally be read as UCU or CUC. They also found that synthetic messenger RNA containing the sequence AAGAAGAAGAAG . . . led to the formation of three polypeptides, each containing only one amino acid–lysine, arginine, or glutamic acid. Again, the idea of a triplet code explained these results, for the sequence of bases in the RNA could be read in three ways: AAG, AGA, or GAA.

Taken together, these results proved that the genetic code is triplet, nonoverlapping, and read in sequence without gaps. The findings were announced by the Wisconsin researchers at the annual meeting of the Federation of American Societies for Experimental Biology in 1965.

By making use of all the work on codon spelling, whether with artificial messenger RNA or the stimulation of the binding of transfer RNA-amino acid complexes to fragments of messenger RNA at the ribosomes, a list of codon assignments could be drawn up for every amino acid. (See end of this chapter.) In nearly every case, the first two letters were the same in the codons for any particular amino acid. It looks as though they usually suffice to identify the site for transfer RNA.

WHY A PARTICULAR CODON?

If the function of transfer RNA is correctly envisaged, the transfer RNA for each amino acid will link up in a specific way with that amino acid at a recognition site of some sort. Only in this way can a particular transfer RNA team with its own amino acid. But what decides which is the amino acid appropriate to it?

A clue to this problem was provided by Marshall Nirenberg in 1965, when he noted that amino acids with similar structures often have similar codons. Stephen R. Pelc at King's College, London, began to wonder whether the specific teaming of transfer RNA and amino acid might come from the fitting together of the three-dimensional shapes of amino acid and codon.

He therefore began by setting out the amino acids in relation to each codon, matching them against only the first two letters, since the third seems to be of subsidiary importance. From this he found that his hunch was confirmed: the more alike amino acids are to each other structurally, the more alike are their codons.

Going further, he noted that when only a simple chemical rearrangement was required to change one amino acid to another, their corresponding codons were only one letter different from each other. When there was a greater difference between amino acids, the respective codons were usually different in both the first and second letters.

Along with his colleague Margaret G. E. Welton, Pelc now began building models of the codons and the relevant amino acids, testing how well they fitted together. Because the third base in the codon seems relatively unimportant, they assumed that this base fitted to an NH_2.CH.COOH group which is found in every amino acid and so is not specific to any one.

By juggling with the rest of the amino acid model, the two King's College workers tested whether it could fit to the other two bases in the codon. They concluded that it could and after working through the twenty common amino acids they announced early in 1966 that nearly all of them could be matched up with their codons in this way.

About a year later Francis Crick checked over some of the models and discovered that they were chemically inaccurate. "I conclude," he wrote in *Nature* in February, 1967, "that the

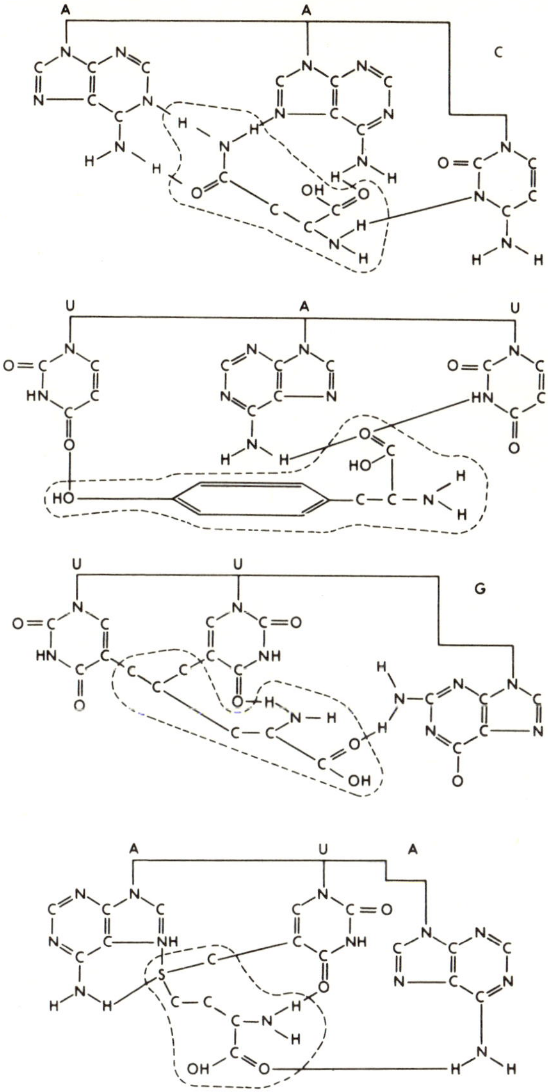

Figure 27. The fit of an amino acid to its codon on transfer RNA as envisaged by Pelc. From top to bottom: asparagine-AAC; tyrosine-UAU; leucine-UUG; methionine-AUA.

models of Pelc and Welton do not support their hypothesis." In an additional note that must be unprecedented in scientific history Sir John Randall, director of the laboratory in which Pelc and Welton were working, stated not only that he agreed with Crick but that he "had so informed Dr Pelc and Miss Welton at the time of their original publication." Attractive as Pelc's idea is, model building has thus far failed to substantiate it.

CODE DEGENERACY

The fact that most amino acids can be bound on to messenger RNA at the ribosome by several codons, often differing only in their last letter, suggests that the binding between the codon on the messenger RNA and the corresponding anticodon on the transfer RNA may depend particularly on the first two bases. The third bases would then pair relatively loosely. Such a possibility, which would seem to explain the general nature of the degeneracy of the genetic code, has been termed the wobble hypothesis by Francis Crick at Cambridge.

As outlined in 1966 by Crick, the wobble rule is remarkably simple. The loose fit requires at least two hydrogen bonds between the third pair of bases and must be reasonably good; apart from that, any pairing of bases is possible.

Thus, according to the wobble theory the base U on the anticodon can pair with A or G in the third place on the codon, the anticodon C can only pair with G in third place, A can only pair with U, G can pair with U or C, and hypoxanthine (another base in transfer RNA) can pair with U, C, or A in third place.

In this connection Crick has suggested that the unusual base inosine is significant. This base, which was found by Holley in the transfer RNA for alanine from yeast, has also been discovered in the transfer RNAs for serine and valine from the same source. Holley has pointed out that IGC may be the anticodon for alanine, and Crick has followed this up by suggesting that the anticodons for serine and valine may be IGA and IAC respectively.

What exact role inosine has is not yet clear, however. For one thing, it does not occur in every variety of transfer RNA. According to Crick, inosine may "recognize" U, C, and A on the codon but not G. In the yeast transfer RNA for tyrosine, which has no inosine, Crick has suggested that the anticodon is GΨA, where Ψ stands for a pseudo-uracil base. Since this can form the same base pairs as U itself, such an anticodon would not conflict with previous data.

In any event, Crick has stated that the wobble hypothesis will very probably explain the degeneracy of the genetic code so far as the choice between U and C in the third base position is concerned. Whether all the code degeneracy can be solely accounted for by wobble is less certain.

Experimental work announced in 1966 by D. Söll, H. G. Khorana, and their associates at Wisconsin supported Crick's hypothesis without going so far as to prove it. Using various transfer RNAs that link up with specific amino acids only, they tested how these would bind onto different codons at the ribosomes, using the technique pioneered by Nirenberg in working out the correct spelling of individual codons. They discovered that even though a particular type of transfer RNA will recognize only specific first two "letters" in a codon, it may tolerate several different third "letters." So far, the variety of third "letters" known to be tolerated is within that prophesied by the wobble hypothesis.

PUNCTUATING THE MESSAGE

The fact that the genetic code is read from a fixed starting point insures that the correct triplet sequence is maintained, but it does not explain how the long DNA molecule is divided into separate sequences, one for each protein. There is now evidence that this punctuation of the genetic message is achieved by means of nonsense codons, which break a sequence and so stop protein synthesis.

In 1965 Sydney Brenner, Anthony Oliver Ward Stretton, and Samuel Kaplan at the University of Cambridge in England, and Martin G. Weigert and Alan Garen at Yale, simultaneously announced that they had found nonsense codons. Both groups of researchers used chemicals that produced changes in codons and studied the proteins that were synthesized as a result. They found that some base sequences resulted in the formation of stunted proteins which ceased to perform their function as enzymes.

The Cambridge group discovered two codons which had this effect: UAG and UAA. They considered that the UAG codon was probably abnormal, but that the UAA codon could well play the role of a period—at any rate, in the bacteria used in the experiments. Nonsense codons do not seem to be universal in the way in which we believe codons for specific amino acids are. The Cambridge workers found that a codon which was nonsense in one bacterial strain made sense in another.

How would such a nonsense codon function as a period? Brenner and his co-workers have suggested that these codons in messenger RNA are recognized by transfer RNAs which, instead of carrying an ordinary amino acid, carry some special compound that terminates the growing polypeptide chain.

Interestingly, at least one antibiotic works in this fashion.

Puromycin contains a nucleotide group and an amino acid group and thus can mimic the action of a transfer RNA-amino acid complex. It takes effect by joining on to messenger RNA at the ribosome and breaking the growth of the polypeptide chain at that point.

STARTING THE POLYPEPTIDE CHAIN

A message must have a beginning as well as an end. What is the beginning of a polypeptide chain that is being assembled on a ribosome? Evidence is appearing, notably as a result of work by Hans Noll at Northwestern University, which indicates that a specially adapted version of the amino acid methionine acts as a "capital letter" in this respect.

Noll has postulated that chain growth proceeds as follows. First, one amino acid (call it x) and its transfer RNA are bound to a site on the ribosome designated the α site. This means that it fits alongside another (y) amino acid-transfer RNA complex already in position on a second site, the β site. The x and y amino acids then link up, with the help of an enzyme called peptide synthetase, to form part of the growing polypeptide chain. The transfer RNA for amino acid y now separates from its amino acid, while remaining bound at site β. When the next link in the chain is forged, amino acid x with its transfer RNA moves to site β with the help of another enzyme, translocase. At the same time this displaces the unattached transfer RNA previously belonging to amino acid y, and this moves into an exit site, γ, where it is loosely bound and from which it eventually leaves the ribosome.

If this is a correct hypothesis, the next problem is to apply it to the actual initiation of a chain. Normally, when an amino acid joins a growing polypeptide chain, it links up with the –COOH group of the amino acid ahead of it on the chain

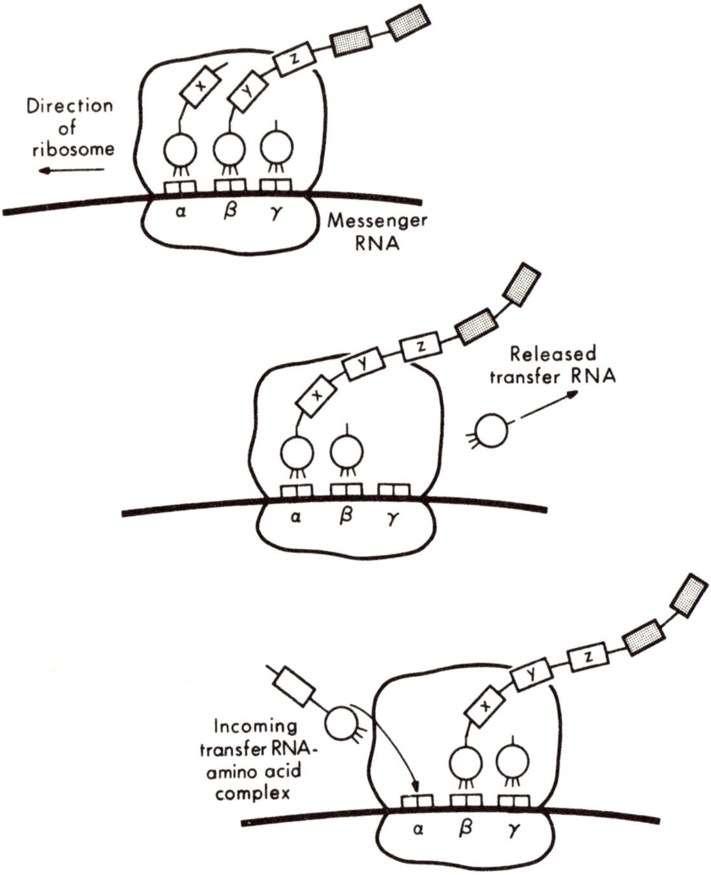

Figure 28. Chain growth at the ribosome may involve three sites on the ribosome, designated α, β, and γ. A freshly arrived transfer RNA-amino acid complex is positioned at site α. After its amino acid has joined the growing polypeptide chain, the transfer RNA is moved to site β, allowing another transfer RNA-amino acid complex to occupy site α. When the amino acid from this second complex in turn joins the chain, the transfer RNA at β is disengaged from its amino acid and moves to exit site γ.

through a positively charged version of its $-NH_2$ group, $-NH_3^+$. When joined to the $-COOH$ group, this becomes an uncharged $-NH$ group. If the translocase enzyme normally shifts the growing polypeptide chain along the ribosome as just described, it must be able to recognize this uncharged $-NH$ group in the peptide chain-transfer RNA complex. However, for the start of the chain there will be no such group, only the $-NH_3^+$ group of the first amino acid. Considering the specificity of enzyme action, it is hard to believe that translocase can equally well recognize this. A similar problem exists with regard to the operation of peptide synthetase. Yet if these enzymes cannot work, how will chain growth ever proceed beyond the first amino acid?

A possible answer favored by Noll is that the first amino acid has its $-NH_2$ group specially adapted by a chemical change called formylation. This involves the addition of

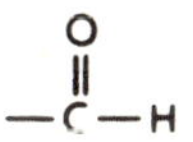

so that the end group becomes

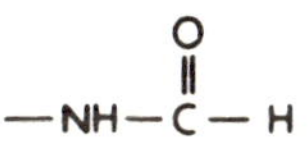

The effect of this would be to mask the existing group and simulate the $-NH$ group. Possibly, this would allow recognition by the enzymes involved.

One amino acid, methionine, can be formylated in this way after it has been linked to transfer RNA. Studies on the amount of binding of this complex to ribosomes strongly suggest that formylated methionine will initiate polypeptide chain formation. If this is so, methionine acts as a "capital letter."

THE GENETIC CODE

First position	Second position U	Second position C	Second position A	Second position G	Third position
U	Phe	Ser	Tyr	Cys	U
	Phe	Ser	Tyr	Cys	C
	Leu	Ser	normal period	Tryp	A
	Leu	Ser	less common period	Tryp	G
C	Leu	Pro	His	Arg	U
	Leu	Pro	His	Arg	C
	Leu	Pro	Glun	Arg	A
	Leu	Pro	Glun	Arg	G
A	Ileu	Thr	Aspn	Ser	U
	Ileu	Thr	Aspn	Ser	C
	Ileu?	Thr	Lys	Arg	A
	Met (capital letter)	Thr	Lys	Arg	G
G	Val	Ala	Asp	Gly	U
	Val	Ala	Asp	Gly	C
	Val	Ala	Glu	Gly	A
	Val	Ala	Glu	Gly	G

The left-hand column lists the four bases representing the first letter of the codon; the second letter is one of the four bases listed across the top of the second column; and the third letter is one of the bases listed vertically in the third column.

It seems possible that in fact methionine is a universal chain initiator, but that in some chains it is removed after the chain has been released from the ribosome. Certainly, not every polypeptide chain, or every protein, has methionine at one end. Another possibility is that several amino acids can act as capital letters, but this hinges upon whether they can be formylated in the necessary manner. So far, we do not know whether they can.

Special starting codons may also be required for methionine and any other amino acids involved as chain initiators. Present evidence suggests that such codons would not be those regularly coding for these amino acids. Thus UUG, normally coding for leucine, has been found to code for formylated methionine in chain initiation. It looks, therefore, as though a codon may have different meanings depending on whether it is coding for chain initiation or for amino acid addition later in chain growth.

11

Genes in Action

UNDERSTANDING OF THE GENETIC CODE MARKS THE END OF AN era in molecular biology. The most immediate problem today is even more complex: what governs whether a gene acts or not?

Every cell in the body of an animal or the structure of a plant contains the same complement of chromosomes. Each of these chromosomes carries an array of genes made out of DNA, and we believe that this allocation of DNA is identical in each cell. Yet the cells themselves vary greatly.

From the moment of fertilization and the first division of the zygote into daughter cells, some kind of distinction in cell function is destined. By the time the embryo has formed its first organs, a tremendous differentiation of cells has occurred. In the human embryo, for instance, cells differ depending on whether they belong to the nervous, muscular, digestive, excretory, circulatory, or respiratory systems.

Further, each type of cell uses its genes to a varying extent at different times, depending on its needs at the moment.

Thus, a liver cell may produce a plentiful supply of one enzyme at one time and then switch to producing another later on.

The set of genes in each cell seems comparable to the range of notes on a piano. Many chords can be selected and played out of the range of notes, and many combinations of genes can be selected and employed out of the set in the cell. The genes which are not used are still there; we say that they are repressed.

FEEDBACK CONTROL

What governs whether a gene is repressed or not? One immediate clue comes from a very widespread method of biological control known as feedback. This describes the way in which a living system stays adjusted to its environment because of some self-regulating control mechanism.

A familiar example of a self-regulating system is a thermostat. If we set a refrigerator thermostat at a certain value, the refrigerator motor starts up as soon as the temperature inside the refrigerator rises above that value.

Similarly when we run, the heart speeds up because our muscles are using up more oxygen than usual and signal that they need a better blood supply to bring them fresh oxygen more quickly. If we eat sweet or starchy food, this is digested into sugar molecules which are carried around in the bloodstream. This rise in blood sugar triggers the pancreas to increase its output of insulin, a hormone needed in the metabolism of sugar.

These are feedback systems in the body as a whole. In the individual cell there are other feedback control systems which govern the functioning of the enzymes. Jacques Monod and Germaine Cohen-Bazire of the Pasteur Institute first discovered that a cell's synthesis of a particular enzyme is controlled by the presence of the substance which the enzyme produces.

The particular enzyme the two French researchers were studying was tryptophan synthetase, which is responsible for synthesizing the amino acid tryptophan. They found that when *Escherichia coli* bacteria were growing in a medium containing plenty of tryptophan, they stopped making tryptophan synthetase.

This type of control, in which the presence of the product helps stop the machinery for making it, is termed negative feedback.

A more positive control is also needed in the cell, so that it can begin to make some substance in response to a new need. The induction of enzyme synthesis for this reason was discovered at the beginning of the twentieth century by Frédéric Dienert of the Agronomical Institute in France.

Dienert was studying the way in which yeast cells ferment lactose, the sugar found in milk. He found that when strains of yeast had been grown in a medium containing lactose for several generations, their enzyme machinery was geared to dealing with this sugar and, once given lactose, they would start fermenting it within an hour. By contrast, yeast cells that had not been grown in lactose were not ready to deal with it and when it was offered to them they could not at first ferment it. However, they adapted by synthesizing the necessary enzyme and after fourteen hours began fermenting the sugar. Evidently the presence of lactose had triggered the synthesis of the specific enzyme that attacks it.

GENES TO CONTROL GENES

Before an enzyme can be synthesized, the gene which controls its formation must go into action and send out instructions in messenger RNA. After this, synthesis of the enzyme on the ribosomes will follow automatically according to the specifications embodied in the messenger RNA. But how is the action of the gene switched on or off according to the cell's needs?

Some brilliant work to answer this question was announced in 1961 by François Jacob and Jacques Monod at the Pasteur Institute. They had been studying how gene mutations affect the capability of *Escherichia coli* bacteria to produce enzymes.

When offered a suitable sugar these bacteria, like yeast cells, manufacture the necessary enzymes to attack the sugar and break it apart. The two French researchers focused their attention on the bacteria's production of β-galactosidase, which catalyzes the breakdown of lactose sugar.

They found that one kind of gene mutation will stop the bacterial cell from making β-galactosidase under any circumstances. They concluded that the affected gene is responsible for actually making the enzyme and termed it a structural gene.

Another type of mutation destroys the bacterial cell's usual ability to regulate production of β-galactosidase. Instead of making the enzyme to order only, the mutant bacteria pour it out all the time. The French workers concluded that the structural gene was normally under the control of a second, regulator gene. It was the functioning of the regulator gene which had been destroyed by this type of mutation.

Since the mutation of the regulator gene caused nonstop production of the enzyme, Jacob and Monod deduced that the regulator would normally switch the structural gene off by producing some substance to inhibit the structural gene's activity.

Evidence to support their idea came from a highly elegant experiment which they devised and carried out. Previous work by Jacob and another colleague, Elie Wollman, had shown that some strains of *E. coli* can "mate" by passing chromosomal material from the "male" cell to the "female." This material is passed between the cells like a rope.

Working in a crowded attic laboratory, Jacob used his wife's mayonnaise mixer to swirl the mating bacteria around and found that he could break this rope any time he wanted to after mating had begun. Wollman and Jacob were therefore able to study the way in which genes were arranged on the rope, by studying how the female cells were genetically modified after receiving various portions of it.

In this way, Wollman and Jacob discovered that the genes were strung out along the chromosome thread like so many beads on a necklace. Thus it took eighteen minutes for four genes to pass from male to female, twenty-five minutes for five genes, thirty minutes for six genes, and so on.

Using the same method, Jacob and Monod now explored how the regulator gene acts upon the structural gene. They used a female strain of *Escherichia coli* in which both types of genes concerned with the making of β-galactosidase were defective, and a male strain in which both genes were working normally.

To start with, both male and female strains were grown in a medium that had no lactose in it. Then the two strains were mixed together, and mating began. From the previous work on gene location, the researchers knew that both regulator and structural genes passed into the female around eighteen minutes after mating had begun.

Soon after the genes passed into the female cells, these cells began making β-galactosidase, even though no lactose was available! They were acting like mutant cells in which the regular gene was destroyed. About an hour later, however, production of the enzyme was switched off.

The researchers concluded that at this time the regulator gene had gone into action and made some compound which stopped the structural gene from making any more β-galactosidase. For this work, they were awarded a share of the Nobel Prize for Physiology and Medicine in 1965.

Still another type of mutation affects production of β-galactosidase. This third type prevents any switching off of the structural gene even though the regulator gene is intact. Jacob and Monod have concluded that there must be a third type of gene, which they call the operator, that controls the functioning of the structural gene, or of a group of structural genes.

Fine genetic analysis has revealed that the operator gene is right beside the structural gene or genes it controls. It seems that the represser substance which is made by the regulator gene does not act directly upon the structural gene. Instead, it binds itself to the operator gene.

An operator gene together with its structural genes forms a functioning whole which Jacob and Monod have called an operon. In some systems, one regulator gene controls one operon; in others, it controls many operons.

How does this regulation of an operon depend upon the presence of some outside substance? It seems that there are two main systems, the inducible and the repressible.

In the inducible system, the represser substance made by the regulator gene interacts with the outside substance (such as lactose) and so is prevented from reaching the operon and functioning as a represser. Consequently the operon sends out instructions to manufacture the enzyme needed to attack the outside substance. This is a system of positive feedback.

In the repressible system, on the other hand, the represser substance is not effective until it has interacted with the product of the system which the operon controls. For instance, an essential component in the manufacture of the amino acid tryptophan is the enzyme tryptophan synthetase. This enzyme is produced automatically according to instructions from the operon until the represser interacts with some of the tryptophan produced. Then the represser becomes effective, and inhibits the functioning of the operon. Thus the amount of tryptophan produced is regulated by the amount already in existence. This is negative feedback control.

HOW ARE GENES REPRESSED?

For a gene to function, its DNA must allow transcription of the necessary messenger RNA. When the gene is repressed, therefore, transcription of the DNA must somehow be blocked.

Very interesting work by James Bonner of Cal Tech and his colleagues, announced in 1963, has revealed that simple proteins called histones block the transcription of DNA. Significantly, histones are normally associated with DNA in the chromosomes. In fact, back in 1957 Maurice Wilkins and his co-workers at King's College, London, tentatively proposed from X-ray data that these protein molecules may be wound around the DNA in both grooves of the DNA helix. However, this idea has not yet been confirmed.

Bonner and his colleagues isolated chromosomes from pea seedlings, by grinding the cells open, filtering off the cell membranes, and then centrifuging. As the chromosomes were the biggest and heaviest of the cell structures remaining, they sedimented out first.

The researchers found that this chromosome material, or chromatin, would make messenger RNA if it was supplied with the appropriate nucleotides bearing three phosphate groups. The chromatin contained some DNA, of course, plus some RNA polymerase, the enzyme which is responsible for stringing together triphosphate nucleotides. It also contained nearly 40 per cent by weight of histone protein. In fact, there was almost as much histone as DNA.

Now the Cal Tech workers separated this histone from the

DNA of the chromatin by altering the acidity of the surrounding solution. This broke the bonds between the two substances. After this, the DNA was sedimented out in the centrifuge.

The California researchers now had DNA relieved of its burden of histone protein. How efficiently would it function? To find out, they supplied it with RNA polymerase, and gave it the correct supply of nucleotides. They found that it was about ninety times more effective at making messenger RNA than the crude chromatin. Evidently, the histone in the chromatin had acted as a brake on the effectiveness of the DNA.

As a further check, Bonner and his co-workers obtained DNA which was fully linked to histone but purified from the crude chromatin, and tested its effectiveness. Unlike the crude chromatin, this DNA-histone complex would not make messenger RNA at all!

From this they concluded that the only reason chromatin can make messenger RNA is that not all of the DNA it contains is fully bound up with the histone. Here, then, was still more evidence that histone represses the functioning of DNA. Bonner and his colleagues had succeeded in identifying an agent of genetic repression.

Could histone protein alone be the represser substance sent out by the regulator gene? One of the requirements of this substance is that it can be made effective or noneffective by interacting with some specific small molecule such as lactose or tryptophan. Only a protein is likely to have its conformation, and hence its properties, so dramatically altered by such interaction. Hence it looks as though the represser substance must have a protein component.

On the other hand, the represser substance must also be able to focus itself on a particular operator gene—that is, a specific portion of DNA. How can it do this? Evidently it must be able to decode the DNA, which implies that it functions like a single-stranded nucleic acid such as RNA.

Since there are both of these requirements for a represser

substance, it makes sense to suppose that it has both of these components—protein and RNA. So far, however, a represser substance of this type has not been identified.

HORMONES ACTIVATE GENES

Every spring, plants awaken from their winter dormancy. Buds which have been alive but resting suddenly begin to grow. Stems lengthen, leaves uncurl—and soon an adult plant is greeting the spring sunshine. For this annual marvel to happen, a whole bank of genes has been triggered off. But how? What decides that genes in one organ of the plant will be derepressed, while in another organ they stay repressed? How is the effort of growth started up and coordinated?

One well-known method of coordination in plants and animals employs chemical messengers called hormones. Gibberellic acid, which is a plant hormone, will start dormant plants growing any time if we apply it. So will another substance called ethylene chlorhydrin.

The action of ethylene chlorhydrin was studied by James Bonner and Dorothy Tuan in 1964. They discovered that the immediate effect of administering it to dormant potato buds was to start RNA synthesis. Of course, from this follows all the rest—protein synthesis, DNA replication, cell division, and growth of the hitherto dormant buds.

Many hormones—perhaps all—seem to work by derepressing genes that were previously repressed. Thus they cause the synthesis of messenger RNAs not made without them. These in turn lead to the output of enzymes and other proteins.

One important instance of hormone activity occurs in the regulation of insect development. When insects such as the fruit fly, *Drosophila*, are developing larvae, some of the characteristic bands in the giant chromosomes of their salivary glands become swollen. These swellings are known as puffs. Puffing, it turns out, is associated with the manufacture of RNA. It looks, therefore, as though puffing is visible evidence that the genes are in action.

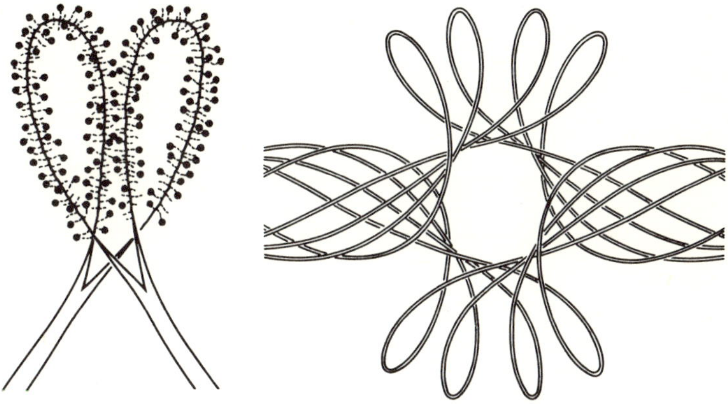

Figure 29. Chromosome puffs as seen under the electron microscope (left) reveal granules believed to be messenger RNA. At right is a schematic representation of how chromosome fibrils untwist to form a puff.

Highly delicate research has shown that the RNAs produced at each chromosome puff are different. This is what we would prophesy if each chromosome band is a separate gene, transcribing its own messenger RNA. In 1963 Walter Sigmund Plaut of the University of Wisconsin did in fact demonstrate that in *Drosophila* each band on the giant salivary chromosomes represents an individual molecule of DNA.

At a chromosomal puff, the DNA appears to straggle out in long loops away from the tight structure of the rest of the chromosome. We would expect this if the DNA were freed of the constricting coils of histone so that it could open out and allow of messenger RNA transcription. Ernest Milton Gifford, Jr., and Herbert B. Tepper at the Davis campus of the University of California found in 1962 that histone disappears from bud cells once the hormone-inducing flowering has reached them. This suggests that hormones may derepress genes by removing histone and allowing their DNA to unwind for messenger RNA synthesis.

To explain why different genes are derepressed in different cells, we need to assume that there are various kinds of histone repressing their DNA. Each kind would respond to a different hormone.

GENES AND MEMORY

Our memory, and indeed our thought generally, must have a physical basis in the brain. Various data suggest that there are two ways in which memory is stored: a short-term process (about half an hour to an hour) and a permanent long-term one. The short-term process may involve some kind of reverberatory electrical activity, whereas the long-term storage is likely to require a structural alteration, perhaps by means of the manufacture of definite molecules.

However, the compounds in the brain are rapidly being broken down and built up again. How can we reconcile this fact with the way in which memory persists? It seems logical to suppose that the permanent factor is a substance that will specify the configuration of new molecules as they are assembled, thus insuring that the memory trace persists. The kind of substance that specifies what molecules are to be manufactured is, of course, RNA.

Various highly significant experiments reveal that in fact there is an undoubted relationship between nervous activity and the manufacture of RNA. For instance, some work in Scandinavia, notably by Holger Hydén at the University of Göteborg in Sweden, has shown that the synthesis of RNA increases when nerve cells in the retina of the eye or the cortex of the brain are stimulated. This would suggest that nervous activity of many kinds, not only memory, leads to the manufacture of RNA.

Going still further, Hydén reported in 1964 that when laboratory rats were learning a new task, the synthesis of RNA in their brain cells was stepped up. On analyzing this RNA, he found that it had its own special composition of bases, as though it had been created to encode the task which the rats were learning. It seemed to be a species of messenger RNA, Hydén reported.

If memory is somehow encoded in messenger RNA, we could expect the memory of animals to be affected if they are given RNA. This may in fact be what happens. Thus in 1963 a team of researchers at the Smith, Kline and French Laboratories in Philadelphia reported experiments which suggested that injecting rats with RNA improved their ability to learn and remember. Also, late in 1965 Allan L. Jacobson and his co-workers in the University of California at Los Angeles reported that RNA produced in the brains of rats that had been trained to do a particular task helped untrained rats to learn the same task! It was as though the UCLA researchers were actually giving the untrained rats memory molecules. However, very careful investigations reported early in 1966 by Marvin Luttges and his co-workers at the Irvine campus of the University of California did not confirm this work. So far, therefore, it has not been definitely established that injection of memory RNA will communicate learning of some specific task.

If RNA's role in memory is to insure the correct synthesis of particular proteins, we should expect that a substance which interferes with protein synthesis would stop the manufacture of memory traces. Experiments by Louis Barkhouse Flexner and Josefa B. Flexner at the University of Pennsylvania were carried out in 1962 to test the correctness of this idea.

These experiments involved checking the effect on learning and memory of puromycin, the antibiotic which stops protein synthesis at the ribosomes. The Pennsylvania researchers injected the antibiotic under the skin of laboratory rats and could find no apparent effect. The next year they tried injecting it directly into the brain. This time, they found that puromycin definitely destroyed memory traces. They even found that different parts of the brain were involved, depending on whether the memory was of recent events or ones further in the past. Upon recovery, the animals tested were all able to begin learning again. Initially, they interpreted this as meaning that the puromycin had acted by inhibiting protein synthesis. To check this idea, they tried other agents known to stop protein synthesis, but found that these had no effect on memory. They therefore assumed that puromycin acts by altering the synthesis of messenger RNAs, which in turn would alter the rate of synthesis of memory proteins. These proteins, they suggested, might operate by affecting the transmission of nerve impulses at the nerve junctions, or synapses. Loss of this messenger RNA would then lead to loss of essential protein and hence loss of memory.

Significant research on a drug which speeds up RNA synthesis was reported to the American Association for the Advancement of Science at the end of 1965. Alvin Jerald Glasky of Abbott Laboratories and Lionel N. Simon from the Illinois State Pediatric Institute told the meeting that the drug, magnesium pemoline, would step up the rate of synthesis of brain RNA both in the test tube and in live rats. It seemed to work by enhancing the efficiency of RNA polymerase and resulted in doubling or trebling the output of RNA.

Would this same drug affect learning and memory? Another researcher at the Abbott Laboratories, Nicholas Peter Plotnikoff, gave the drug to rats in their food and found that they learned faster and remembered longer.

Just what relationship there is between the known activity

of this drug and the way in which memory works is still unclear. However, human trials of the drug were started in 1965, and if they are promising it seems safe to prophesy that college students will soon be pepping up their memories before examinations by dosing themselves with magnesium pemoline.

All of this research on RNA and memory must eventually lead back to the role of DNA in the genes, for this ultimately governs what RNA can be synthesized in the cell. Further work may reveal that our memory is more closely dependent upon our genetic endowment than could ever have been guessed before the days of molecular biology.

MAKING GENES TO ORDER

What may be the end result of our knowledge of gene functioning? Will we be able to use it for our direct benefit?

One scientist who believes that we probably will is Nobel Prizewinner Edward L. Tatum. In his address on receiving the prize in 1958 he made some exciting and thought-provoking speculations on this subject.

> It does not seem unrealistic [he said] to expect that, as more is learned about control of cell machinery and heredity, we will see the complete conquest of many of man's ills, including hereditary defects in metabolism and the currently more obscure conditions such as cancer and the degenerative diseases, just as diseases of bacterial and viral etiology are now being conquered.

If this much control were to become feasible, Tatum argued, men might well be able to go still further and improve the whole functioning of cell processes. He termed this "biological engineering."

How would such biological engineering be attempted? Tatum speculated that it might proceed in stages, starting with the test-tube synthesis of better enzymes, then of the corresponding nucleic acid molecules, and finally by the introduction of these nucleic acids into the gene reservoir of the organism. This last would mean "infecting" the sex cells of the organism with custom-built viruses carrying the required nucleic acid. Alternatively, Tatum added, it might be possible to reach the same goal by carefully directed genetic mutations.

Prophecies such as these hint at the immense powers for good or ill which may stem from molecular biology. Our whole outlook on life is likely to change in the years to come, as we become increasingly accustomed to thinking of bodily and mental functioning only in terms of chemistry. The belief that there is some mysterious quality about life which can never be explained in terms of chemistry dies hard, and the relics of vitalism are still with us. However, molecular biologists are proceeding today in the belief that all living processes will be explicable in terms of chemistry.

BIBLIOGRAPHY

Research in molecular biology is proceeding so quickly that the only completely up-to-date sources of information are research journals such as the *Journal of Molecular Biology*, *Nature*, *Proceedings of the National Academy of Sciences* (Washington), and *Science*. Less specialized but highly valuable explanations of current work are available from time to time in *Scientific American*.

For further general reading the following books are to be recommended:

ANFINSEN, CHRISTIAN B. *The Molecular Basis of Evolution*. New York: John Wiley, 1959.

BARRY, J. M. *Molecular Biology: Genes and the Chemical Control of Living Cells*. Englewood Cliffs, N.J.: Prentice-Hall, 1964.

BONNER, DAVID M., and STANLEY E. MILLS. *Heredity*. 2nd ed. Englewood Cliffs, N.J.: Prentice-Hall, 1964.

LEHRMAN, ROBERT L. *The Reproduction of Life*. New York: Basic Books, 1964.

LEVINE, R. P. *Genetics*. New York: Holt, Rinehart and Winston, 1962.

MCELROY, WILLIAM D. *Cell Physiology and Biochemistry*. 2nd ed. Englewood Cliffs, N.J.: Prentice-Hall, 1964.

RAMSEY, J. A. *The Experimental Basis of Modern Biology*. London: Cambridge University Press, 1965.

STAHL, FRANKLIN W. *The Mechanics of Inheritance*. Englewood Cliffs, N.J.: Prentice-Hall, 1964.

SWANSON, CARL P. *The Cell*. 2nd ed. Englewood Cliffs, N.J.: Prentice-Hall, 1964.

WALLACE, BRUCE. *Chromosomes, Giant Molecules and Evolution*. New York: W. W. Norton, 1966.

WINCHESTER, A. M. *Heredity: An Introduction to Genetics*. New York: Barnes and Noble, 1961.

INDEX